# Industrial Photoinitiators

## A Technical Guide

# Industrial Photoinitiators

## A Technical Guide

W. Arthur Green

CRC Press
Taylor & Francis Group
Boca Raton London New York

CRC Press is an imprint of the
Taylor & Francis Group, an **informa** business

CRC Press
Taylor & Francis Group
6000 Broken Sound Parkway NW, Suite 300
Boca Raton, FL 33487-2742

© 2010 by Taylor and Francis Group, LLC
CRC Press is an imprint of Taylor & Francis Group, an Informa business

Printed in the United States of America on acid-free paper
10 9 8 7 6 5 4 3 2 1

International Standard Book Number: 978-1-4398-2745-1 (Paperback)

### Library of Congress Cataloging-in-Publication Data

Green, W. Arthur.
   Industrial photoinitiators : a technical guide / W. Arthur Green.
      p. cm.
   Includes bibliographical references and index.
   ISBN 978-1-4398-2745-1 (paperback : alk. paper)
   1. Radiation curing. 2. Photochemistry--Industrial applications. I. Title.

TP156.C8G74 2010
660'.28426--dc22
                                              2009054000

**Visit the Taylor & Francis Web site at**
**http://www.taylorandfrancis.com**

**and the CRC Press Web site at**
**http://www.crcpress.com**

*For Peter N. Green … a pioneer.*

*"Pass it on, boys, pass it on."*

**The History Boys**
*Alan Bennett*

# Contents

# Acknowledgments

I joined Ward Blenkinsop & Co. Ltd., Fine Chemical Manufacturers near Liverpool, United Kingdom, in 1969 with a degree in industrial chemistry. My first years were spent as a development chemist in the Pilot Plant, scaling up new processes. In the early 1970s the techniques and equipment that were used were very basic and much of this work was "boys' own, hands-on" chemistry with a vengeance, but this was also tremendous experience and involved a great intimacy with the reactions undertaken. This experience led to a position as plant manager in a small-scale unit producing a wide variety of chemical intermediates and specialties. In 1984, I took an opportunity within the company to join a very small team engaged in product research. Novel photoinitiators were being developed and the company had already introduced several new materials to the developing field of UV curing, including isopropylthioxanthone on a commercial scale.

We continued to develop, test, market, and supply novel photoinitiators, including a huge amount of work on water soluble benzophenones and thioxanthones. During this period we developed a long-term industrial/academic partnership with Manchester Metropolitan University, and I owe Professor Norman S. Allen a huge debt for providing me with some understanding of the minutiae of academic studies on these materials through several PhD studentships, where his students likewise obtained some valuable industrial expertise.

We forged relationships with several companies and research establishments throughout the United Kingdom and internationally, and the constant visits, flow of data, testing samples, and gaining feedback led to many friendships. Patents were filed, technical papers were published, and many presentations were made on several continents, mostly via the RadTech or similar organizations. I have many people to thank for some very interesting and happy times.

There are also those at the company, which later became Great Lakes Fine Chemicals, who gave valuable support and were tremendous workmates. In particular, I would like to thank Dr. Peter N. Green, my line manager in product research, whose enthusiastic encouragement and direction gave me a whole new career in chemistry and who is still a valued friend.

I am grateful for the very welcome comments and advice regarding this book, as well as the significant contribution from Dr. Shaun Herlihy of Sun Chemical Ltd., St. Mary Cray, UK. Much of the information in the book is of a practical nature and stems from the many conference proceedings of the RadTech organizations of North America, Europe, and Asia, as well as those of the Paint Research Association in the United Kingdom. I am grateful for their kind permission to use this material.

Finally, my family have given me much encouragement and help in the writing of this book. Thanks go to Jenny, Andy, Nicki, and Stu for guiding me around my laptop, preparing diagrams, proofreading, and continual nudges in the right direction. Most of all, my wife Cynthia deserves a medal for her endless patience and for allowing considerable slippage in the DIY department.

**Arthur Green**

# Introduction

The ultraviolet (UV) curing industry, using the energy of UV light in the formation of polymeric materials, is barely 50 years old, yet is only now approaching some degree of maturity. The development of monomers, oligomers, and photoinitiators during this time has allowed the technology to advance into very efficient formulations that can service a wide variety of applications.

The use of a photoinitiator to generate free radicals or protonic acids via a UV source of energy is the essence of the UV curing process. This low-energy, cold cure, environmentally friendly technology has made steady inroads into conventional solvent-based resin technology and is now used in coatings, varnishes, graphic arts, high-speed printing, metal decorating, adhesives, laminates, printed circuit boards, imaging, dental, cosmetics, and biocompatible processes—and the list goes on.

The commercial use of photosensitive chemicals and the possibilities of UV curing began to be explored in the 1940s when patents for UV-curable inks and coatings first appeared. By the 1960s, the technology had been proven and production lines for UV coatings and inks appeared. At that point there were few specific photoinitiators available in commercial quantities, and many materials that were used in other industries, such as benzophenone, benzil, and anthraquinones, were applied in the UV industry. Diazo compounds, phosphates, and chromium salts, etc. were used for light-sensitive screen stencils and printing plates.

In the 1970s, specialist chemical manufacturers began to make photoinitiators and benzoin ethers, benzil ketals (Union Carbide, BASF), hydroxyacetophenones (Merck, Ciba), substituted benzophenones and thioxanthones (Ward Blenkinsop), etc. were introduced to the market. This gave UV formulators a much wider choice of photoinitiators and the technology leapt forward. Development continued into the 1980s with Ciba introducing high-speed alkylaminoacetophenones (Irgacure 907) and BASF developing phosphine oxides (TPO), which provided the perfect solution to curing titanium dioxide whites. Variations on all themes continued to appear in the 1990s.

Today, developments in photoinitiators are addressing more specific problem areas such as polymeric photoinitiators for low-migration food packaging inks and specialist areas such as resists and imaging with latent photo-acid generators, and latent photo-base generators for speciality coatings and adhesives. Both academia and industry are continuing to look at new materials, new processes, and new applications in which UV curing can make a contribution.

This book follows the UV curing process from the use of various UV light sources, the mechanisms of the production of free radicals and protonic acids from the photoinitiator, through to the polymerization process involving acrylate and epoxy resins. Commercially available photointiators are examined in some detail—the different types, how they work, their use, and specific applications, etc. Both free radical and cationic photoinitiators are discussed, along with tertiary amine hydrogen donors. Factors affecting the use of photoinitiators, such as the effects of pigments on cure, surface cure, depth cure, the influence of oxygen, etc. are discussed. A chapter on academic studies introduces the detailed background knowledge that is used in the development of novel photoinitiators.

The wide range of photoinitiators now in use means that energy can be absorbed throughout the UV and near-visible spectrum, and this allows the more specific formulation of products for numerous applications. Photoinitiators can be sourced from many outlets and have become almost high-volume commodity chemicals. Individual photoinitiators are sold under a variety of trademarks and labels and are often available from resin suppliers as well as from specific agencies and manufacturers. The main suppliers and trademarks are listed.

Tables list the various physicochemical and photochemical data about the photoinitiators.

Many textbooks are available that discuss photoinitiators and UV curing in much more detail, often of an academic nature, and some of these are listed in Further Reading.

This small volume is not a treatise but is intended for those new to the UV curing industry and those requiring a simple reference book for commercial photoiniators. It explains the theory, practice, some of the problems, and the use of photoinitiators in simple language.

Numbers in parentheses in the text, e.g., (27), refer to the number of the photoinitiator in the Appendix A tables. The references are mainly directed toward practically oriented papers, most of which come from the conference proceedings of the RadTech and PRA organizations.

*chapter one*

# Let there be light

The first and essential step in the ultraviolet (UV) curing process is the absorption of UV energy in the form of light. The photoinitiator that is introduced into the formulation is designed to do this and will respond to photons or packets of light energy to produce a reactive species. The free radical, or the cationic species, that is produced can then initiate a polymerization process depending on the chemistry of the monomers and oligomers that are used. The photoinitiator must be able to respond to the wavelength of the light that is supplied, and matching the absorbance of the initiator to the type of light source, UV or visible, is important for the whole process to take place efficiently.

This chapter examines the nature of UV light and the types of commercial UV lamps that are available. It discusses the light absorption process and the excited states that are produced by the photoinitiator prior to the production of radicals.

## 1.1   The electromagnetic spectrum

The electromagnetic spectrum comprises the broad band of energy that emanates from the sun and includes radio waves, microwaves, infrared radiation, the various colors of the visible spectrum, UV light, x-rays, and cosmic rays.[1] These bands of energy can be described in waveform and are best defined by their wavelengths, most practically for the UV spectrum in nanometers (nm) or $10^{-9}$ meters (one billionth of a meter).

Most of these wavebands can be generated artificially and can be used for a multitude of purposes. The longer wavebands are used for telecommunications, radio, etc. ($10^{-1}$ to $10^3$ m) and microwaves ($10^{-2}$ m) can be used to initiate chemical reactions or as an indirect heat source in suitable ovens. At shorter wavelengths, the infrared, from 700 to 2000 nm, can provide energy in the form of radiant heat, and chemical initiators are available that can also respond to these wavelengths.

The visible spectrum provides the colors from red light at 700 nm, through the rainbow of yellow and green to blue light at 400 nm. Many lasers are available that can provide light of a discrete wavelength in the visible spectrum, such as the Neodynium-YAG laser at 430 nm, the Argon laser at 488 nm, the Krypton laser at 647 nm, and the Ruby laser at 694 nm.

*Table 1.1* The Electromagnetic Spectrum and the UV Spectrum
(Wavelength λ in Nanometers)

| Cosmic Rays | X-rays | UV | Visible | Infrared | Microwaves | Radio |
|---|---|---|---|---|---|---|
| 0.001 | 0.1–10 | 100–400 | 400–700 | 700–2000 | 1 million | 1 billion |

| | | | | Visible | | | |
|---|---|---|---|---|---|---|---|
| **Deep UV** | **UV C** | **UV B** | **UV A** | **Blue** | **Green** | **Yellow** | **Red** |
| 100–200 | 200–280 | 280–320 | 320–400 | 400 | | | 700 |

The small band of visible light at 400–450 nm that can activate some commonly used photoinitiators is also known as the near-visible, UV-VIS, or UVV, and sometimes as blue light.

At shorter wavelengths, the UV spectrum covers the energy band from 100–400 nm. The UV spectrum is often described as UV A, UV B, or UV C and these refer to the respective wavelength bands of 320–400 nm, 280–320 nm, and 200–280 nm. Vacuum UV or Deep UV refers to the band of UV energy covering 100–200 nm. UV A, B, and C may also be referred to as long-wave UV, mid UV, and short-wave UV.

Photons emitted at short wavelengths have higher energy levels than those at longer wavelengths, and are less penetrative regarding the depth of cure that can be achieved. UV light at 400 nm will penetrate much deeper than that at 250 nm.

At even shorter wavelengths there is the region of very-high-energy x-rays ($10^{-12}$ m). These different wavelength bands, shown in Table 1.1, do not have definitive cut-off points; the UV and visible spectrum is a continuous band of energy transmitted by photons or "packets" of light. The categories are simply of a descriptive nature.

The majority of photoinitiators in common use by the UV curing industry can be activated by the energy provided by the UV and near-visible spectrum, from 200 nm to 450 nm, and it is irradiation sources, or light, at these wavelengths that are generally used.

## 1.2   The generation of UV energy

The UV curing industry has been using mercury lamps to generate UV light since its inception, and the medium-pressure mercury lamp (MPM lamp) has been developed to fulfill almost all demands put upon it since then. In more recent times, new types of lamps have appeared that are either more efficient or more suited for a particular purpose, such as flash lamps for UV screen inks and the growing development of UV LED lamps.

The different types of lamps that are used to provide UV and UV-VIS light are described as follows.[2]

## 1.2.1   The medium-pressure mercury lamp

The most widely used UV lamp in the industry is the MPM lamp, sometimes called the "H" lamp. These lamps are available in a variety of lengths, diameters, and power ratings to suit most press applications.

When mercury vapor is excited by an energy source, the electrons in the mercury atom are promoted to several higher-energy levels. Since these excited states are essentially unstable, they lose their energy and the electrons revert to their original ground state orbitals. Energy is released as photons during this latter process and takes the form of discrete bands of both UV and visible light. Mercury is particularly useful since it is readily vaporized and provides a wide range of emission lines throughout the UV spectrum.

The MPM lamp can be activated either by a high voltage electric arc (in the standard model) or by microwave power (in the Fusion model).[3] In both cases, the five main bands of UV light that are produced are at discrete wavelengths of 254 nm, 313 nm, 366 nm (i-line), 404 nm (h-line) and 436 nm (g-line). There are also smaller outputs at other wavelengths including a good deal of visible light and infrared (IR).

### 1.2.1.1   The standard mercury arc lamp

The standard mercury arc lamp, illustrated in Figure 1.1, is composed of a sealed quartz tube with tungsten electrodes at each end. The tube contains a fixed amount of mercury to provide one atmosphere pressure of mercury vapor while under working conditions plus a small amount of a starter gas, normally argon.

The low pressure starter gas is easily ionized and allows current to flow and heat up the lamp to the operating temperature of 600–800°C, at which point all the mercury is ionized and an arc is struck. It can take a few minutes to form a stable arc and reach full power. When the lamp is switched off, the mercury ions and electrons that support the arc combine

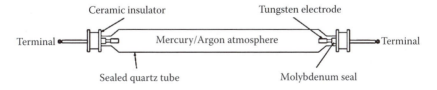

*Figure 1.1* The medium-pressure mercury arc lamp.

and the mercury vapor condenses. Immediate restart is difficult with a hot lamp and time must be allowed for it to cool down before the start-up process can be repeated. Electrodes slowly deteriorate by sputtering during start-up as the arc is formed, causing small amounts of tungsten to be deposited on the tube walls at each end. This steady electrode decay during start-up is usually the determining factor in the lifetime of a lamp. To prolong lamp life, it is therefore necessary to reduce the need for switching on and off to the minimum. Stand-by power technology can prevent frequent switch-off when a "press down" is needed, and shutters can also reduce overheating of the substrate.

There is a slow decline in UV intensity over time, mainly in the short wave, and it is difficult to judge this deterioration without constant measurement of the lamp output at various wavelengths.

Ballast circuits are used to smooth out power fluctuations and stabilize the lamp arc for a continuous, steady dose rate. Recent developments include "Quick Start" technology that requires no stand-by or shutters on the lamp.

Available lamp power has increased over the years from 80 W/cm to 240 W/cm. Most applications can be run economically on the modern 180 W/cm lamp design, which is now equivalent to the older 240 W/cm. Overall conversion of electrical energy to UV is only around 25%. Higher lamp power comes with the disadvantages of increased electrode wear at start-up, increased heat management, and the requirement for much higher electrical power supplies.

A narrower diameter of lamp is available and this gives a more tightly focused beam. The dose rate is improved by 25%, particularly in the UV C wavelengths at 200–280 nm, which improves surface cure and cure speed.

About 50% of the energy produced by the lamp is in the form of heat, and heat management, coupled with reflector design to focus the UV light, is an important subject.[4] Heat is generated in the form of infrared energy and this can influence the photocuring process. In general terms, the heat that arrives with the UV at the film surface during cure will speed up the curing process. This may help to some extent, particularly in the cationic curing process. However, heat may be detrimental in affecting sensitive substrates, producing distortion and curl in plastics, for example.

Heat is also detrimental for the press and can give rise to some minor distortion as a result of the metal press frame's expanding. This leads to subtle changes in press dimensions that affect print register, and many press manufacturers are keen to explore any options for an improvement on "cold curing" UV sources.

Lamp reflectors, shown in Figure 1.2, come in two basic shapes: the classical designs for focusing or for distributing light or radio, TV signals, etc. Elliptical reflectors focus the UV light to a focal point on the substrate

Elliptical reflector          Parabolic reflector

*Figure 1.2* UV lamp reflector design.

although in practice this is usually just an area of narrow-beam, higher-intensity UV rather than a single line. Focusing quartz lenses may be placed between the lamp and the substrate to achieve very high-intensity, narrow-line UV if required. Parabolic reflectors give a more even, flooded area of UV. In both cases, cooling systems may be incorporated into the design of the lamp housing. Both lamp and press manufacturers tend to have their own particular designs to accommodate heat management.[4]

Most UV lamps come with water-cooled systems and filter units that can remove the IR and provide "cold cure" UV. Quartz water cooling tubes can be placed between the lamp and the substrate. These tend to restrict the generation of ozone but may also remove some short wave UV. Quartz plates cooled by an air flow between the lamp and the substrate can remove IR. Chilled rollers beneath the substrate can also provide some cooling. There are disadvantages with water cooling systems in that deionized water must be used to prevent the loss of too much short wave UV through absorption by contaminants.

The most effective system for removing IR heat is probably the dichroic reflector,[5,6] where multilayer coatings on the reflector allow the IR energy to pass through the coating and be absorbed by the reflector body, whereas the UV light is reflected from the top surface of the coating. Some IR heat will always be present, coming directly from the lamp to the substrate without reflection.

The lamp housing may also include some facility for nitrogen blanketing to improve cure, but this is a complex mechanical problem and the use of nitrogen depends very much on the application and the type of press (mainly web). The economics of using nitrogen, with significant installation and running costs, is not easily determined. Nitrogen blanketing is discussed in more detail in Chapter 5.

### 1.2.1.2   The electrodeless microwave-powered UV lamp

Instead of using an electric arc to excite the mercury atoms, this type of lamp is powered by 2450 MHz microwave energy provided by two 1.5 kW or 3.0 kW magnetrons that feed the energy to the lamp via two waveguides (Figure 1.3).

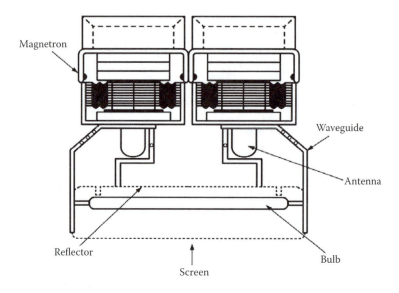

*Figure 1.3* The electrodeless microwave-powered UV lamp.

The lamp is a sealed quartz tube containing mercury and starter gas, very similar to the arc lamp, but without the electrodes. Excitation of the mercury atoms is very fast and the lamps are essentially instant on/off. There is no warm up or cool down period required and heat management is limited to air cooling of the lamp and magnetrons. A wire mesh between the lamp and the substrate prevents any leakage of microwaves.

Lamp size is limited to 15 cm or 25 cm, but the lamps are easily stacked together to provide longer lengths. A smaller bulb diameter can also be used which gives a more focused and higher dose rate of UV compared to the standard size lamp. These narrower lamp dimensions give the microwave-powered lamps a distinct advantage in higher cure speeds due to increased UV C.

Lamp lifetimes in the absence of electrodes are much longer. This also means that doped lamps operate much better under this set up, and many types of doped lamps have been developed for the microwave lamp.

Solid state electrics in the Fusion model lamp can increase the power up to 360 W/cm and provide increased microwave efficiency, but the heat management at this level becomes problematic. The standard mercury arc lamp design is more electrically efficient overall. More recent developments in electrodeless lamps are heading toward miniaturization, using smaller magnetrons or radio frequency excitation for microcircuits and

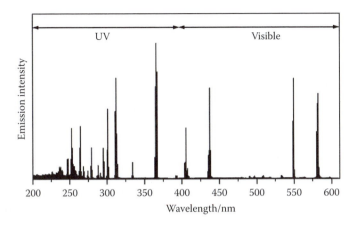

*Figure 1.4* The UV and visible output of a medium-pressure mercury lamp.

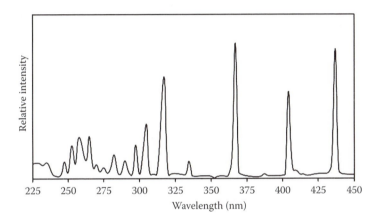

*Figure 1.5* The UV output lines of a standard medium-pressure mercury H lamp.

spot curing, etc. The MPM lamp, from both power sources, provides a fairly polychromatic wavelength distribution. See Figures 1.4 and 1.5.

### 1.2.1.3    Wavelength output of the medium-pressure mercury lamp

Photoinitiators absorb light over a small wavelength range of UV, and the different types of photoinitiators available will provide absorption patterns that cover most areas of the UV spectrum.

The MPM lamp provides a wide range of output lines over most of the UV spectrum and several lines in the visible, some of which are important for UV curing. Figure 1.4 shows the output distribution of the MPM lamp including the visible light. Most commercial photoinitiators will respond to one or more of these output lines and be suitably activated.

Figure 1.5 shows in more detail the UV output of the medium-pressure lamp. The UV C output lines of the MPM lamp between 220 nm and 280 nm are very weak and show little penetrative power due to the strong absorption of the oligomers in this region. They do, however, provide good surface cure, particularly in thin films and varnishes. In this respect, lamp dimensions are important, and a narrow lamp diameter will give up to a 30% increased dose of UV C.

The UV B output lines from 280 nm to 313 nm show increased intensity and better cure characteristics, leading to some useful depth cure as well as surface cure. The UV A and visible output lines at 366 nm, 404 nm, and 436 nm provide increased penetration and are primarily responsible for depth cure where thioxanthones or phosphine oxides are employed as initiators. The strong output lines at 550 nm and 580 nm in the visible range are of no significance in UV curing terms. The main output lines of the mercury lamp for UV curing purposes can be considered as those at 254 nm, 313 nm, 366 nm, 404 nm, and 436 nm. There are moderate outputs in the short wave at 265 nm, 297 nm, and 302 nm.

The power of the UV lamp can be described in several ways.[3] The UV dose, in $mJ/cm^2$, refers to the total light energy irradiated per unit surface area. UV dose is inversely proportional to belt speed and proportional to the number of passes under the lamp. The UV irradiance, in $W/cm^2$, refers to the radiant power arriving at the surface per unit area. Irradiance is a characteristic of the lamp and reflector geometry and does not vary with speed. Peak irradiance is the intense, focused peak directly under the lamp. Intensity is sometimes used as a term instead of irradiance.

## 1.2.2   Doped lamps

The relative output power of the lamp at various wavelengths can be modified by adding traces of other metals, as metal halides, to the mercury to produce "doped" lamps.[7–9] Alterations to the output wavelengths may also occur. Doped lamps are available for both the standard mercury lamp and the electrodeless microwave lamp. The latter provides the more stable form of doped lamp and its use is becoming more common.

The disadvantages of doped lamps are that they run hotter, are more difficult to maintain a stable arc, are more prone to electrode decay in the standard MPM version, and have a reduced lifetime.

The specific UV output spread of a doped lamp slowly changes back to that of the standard mercury lamp with time. This is due to the slow deposition of the dope metal and loss of dope concentration. Most of these disadvantages are being addressed and the use of doped lamps is growing considerably.

Many trace metals have been used, but the most common are iron and gallium. More recent developments have focused on the use of doped lamps (gallium) for UV ink-jet to cure relatively thick layers without the presence of short wave UV that may create ozone in a working environment.

Traces of iron, lead and gallium will increase the UV output at longer wavelengths in the 366 nm and 404 nm region, making these lamps more suitable for curing thick pigmented coatings. Doping with tin produces a different spread of UV output throughout the spectrum. Many more metals have been used including silver, aluminum, bismuth, cobalt, cadmium, indium, magnesium, manganese, nickel, antimony, etc., and more recent developments include blends of dope metals.

The "D" bulb, Figure 1.6, is doped with iron, which increases the output power in the mid to long wave UV and makes this lamp more suitable for inks. Iron-doped lamps can provide increased cure with black or darker-colored screen inks and are particularly beneficial for very thick coatings to achieve good depth cure.

The "V" bulb, Figure 1.7, is doped with gallium and gives a strong output in the near-visible range, suitable for titanium dioxide white coatings and screen inks. The V lamp with practically zero short wave UV is also suitable for use with sensitive substrates such as polycarbonates, which may be degraded by short wave UV. The lack of UV C also means that this lamp will not produce ozone and can be safely used in an "office" environment.

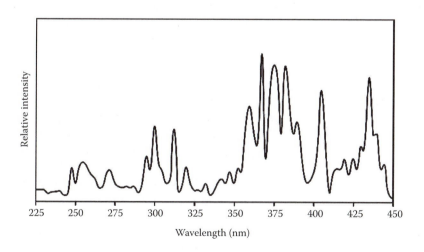

*Figure 1.6* The UV output power distribution of an iron-doped D lamp.

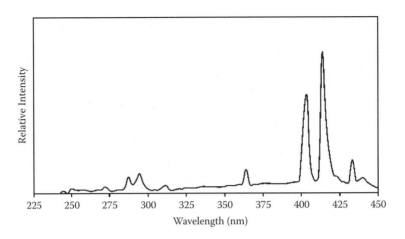

*Figure 1.7* The UV output power distribution of a gallium-doped V lamp.

### 1.2.3   The low-pressure mercury lamp

The low-pressure mercury lamp produces most of its output in the short wave UV range at 185 nm (5%) and 254 nm (95%), with very little output in the rest of the UV spectrum. These lamps are sometimes called germicidal lamps as they have relatively low power (1 W/cm) and are used for applications such as water sterilization. Short wave UV is also used for microchip production and resists. UV C provides limited penetration of light at these short wavelengths and, while they can give good surface cure, these lamps are not widely used for more general UV curing.

### 1.2.4   The high-pressure mercury lamp or capillary lamp

At pressures above 10 atmospheres the spectral output changes from a line structure to a more continuous spectrum. When xenon is included with the mercury, a continuum of light is formed that covers much of the UV spectrum as well as some visible and infrared.[10] The light output is ten times more intense than a medium-pressure lamp and requires water cooling, but these lamps tend to be less efficient in the UV and are not widely used for curing inks and coatings. They are more suitable for near-visible applications such as photolithography.

### 1.2.5   Flash lamps

A xenon tube emits more of a continuum of UV and visible light with little emission below 240 nm, so the production of ozone from short wave

UV is negligible. When a capacitor is discharged through a xenon tube it releases a flash with a huge power density compared with a mercury lamp.[11] These flash lamps can operate very fast, several flashes per second, but the short pulse means that the heat build-up is minimized.

Flash lamps are easily made and can also be shaped to provide circular tubes for CD/DVD curing etc. They have penetrated the screen ink market recently, where easy installation between colors gives partial cure before a final MPM lamp is used. The advantage of screen printing is that it is a stationary printing process where flash illuminates the whole print. Using flash for a moving web would provide bands of uneven cure due to the on–off nature of the flash. Flash lamps can provide excellent cure of thick, pigmented coatings and have been used for curing composites.

## 1.2.6   Fluorescent lamps

Fluorescent lamps are essentially low-pressure mercury lamps with a phosphor coating that converts the short wave UV to visible light. Fluorescent lamps that emit only UV A can be used cosmetically for "tanning" purposes, but they have little relevance to UV curing applications.

## 1.2.7   Excimer lamps

Excimer lamps work by the excitation of certain materials, such as xenon and chlorine, to produce excited dimers or excimers, which revert to their ground state and release energy at a specific wavelength. Xenon emits at 172 nm, krypton chloride at 222 nm, and xenon chloride at 308 nm, the latter being the most useful commercially. Power comes from a dielectric discharge between electrodes in concentric tube design with a water cooled quartz jacket to provide a truly cold-cure lamp. UV light is generated by a large number of micro discharges from the excited dimers. No infrared is generated. The excimer lamp is instant on–off and has a long lifetime.

The early excimer lamps did not have sufficient power to succeed commercially, but more recent developments have increased their power considerably. The single line output at 308 nm means that the choice of photoinitiators that will respond to this is limited and formulation is very restricted. This limited spectral spread and the lower intensity make the excimer lamp unsuitable for more than specific applications.

## 1.2.8   Light emitting diodes

Light emitting diodes (LEDs), being tiny devices that can deliver light in the most difficult of places, are very flexible sources of UV energy. There has been strong development recently in LED technology that can provide power at discrete wavelengths throughout the visible and UV spectra. LEDs are now being used in numerous applications from traffic lights to biomedical sensors.

LEDs emitting UV light can be arranged in SLMs (semi-conductor light matrices) of up to 120 dies, or diodes, per unit, which gives sufficient spread and power for UV curing applications.[12,13] Their advantages are much more efficient energy use compared with mercury lamps, an essentially "cold-cure" source and instant start. The disadvantages that are restricting progress are the low light intensity of the modules, no more than 10 W/cm$^2$ compared with mercury lamps at 180 W/cm$^2$, and the necessary cooling equipment that is required, since the power drops dramatically if the temperature of the LED die rises above 50°C.

Commercial LED arrays are now available that will deliver UV light in a single peak at 395 nm with a very narrow bandwidth. More recent models emit a similar single peak at 365 nm. Combinations of these two wavelengths will give the most flexible use of LEDs regarding formulation (Section 5.1.2). The power output at the surface is very low, typically around 1–2 W/cm$^2$ at 365 nm and 6–9 W/cm$^2$ at 395 nm. Despite this low output, UV formulations have been produced that will give full cure. UV LEDs have also become available that will emit in the short wave UV range and provide light at wavelengths from 255 nm to 355 nm. However, these shorter-wavelength UV LEDs emit even less power and are very expensive and unlikely to become more than research tools.

Full cure using LED lamps will depend on the coating thickness and the effectiveness of both surface and depth cure. In standard UV curing, different wavelengths are used to give full cure, where high concentrations of short wave initiators will give good surface cure and low concentrations of long wave initiators will give depth cure. For LED systems, the single output line, like those of excimer lamps, offers limited formulation possibilities with no shorter wave UV for fast surface cure. In addition, the low output intensity of LEDs limits their use in graphic arts and they can really only be considered specialist devices at the moment, although LED technology is gaining use in UV ink-jet.

### 1.2.9   Lasers

Lasers generate specific wavelengths of light by using combinations of materials in a crystalline form that can "tune" a light source to a particular wavelength. Several types of laser are used for photoimaging and microcircuit work on silicon chips.[14,15] Short wavelength lasers at 172 nm (xenon) or 222 nm (krypton) can be used for surface modification to improve coating reception of particular formulations. Typical lasers include helium-cadmium 325 nm, nitrogen 337 nm, krypton ion 337 nm, argon-ion 363, 488 and 514 nm, neodymium-YAG 532 nm, and helium-neon 633 nm, etc. Lasers can provide much more power than UV or visible LEDs.

## 1.3   The light absorption process

The absorption of UV energy by a photoinitiator results in the occurrence of a series of energetic processes. Initially a high-energy singlet state is produced that may convert to a more stable, but less energetic, triplet state by intersystem crossing. Most photoinitiators produce radicals from the triplet state but a few can interact from the excited singlet state (see Figure 1.8).[16]

There are also several decay processes that may occur from the excited states and the main ones are shown in Figure 1.8. The singlet state may decay to the ground state by fluorescence or convert by intersystem crossing to an excited triplet state. The triplet state may decay by phosphorescence or be quenched by monomers or oxygen, or go on to produce radicals by various chemical mechanisms. The study of fluorescence and phosphorescence that is emitted at different wavelengths from an excited species can provide insight into the electronic transitions that are occurring and their energy levels.

For the generation of free radicals, a photoinitiator must absorb UV energy at as high a rate as possible and produce an excited state. This

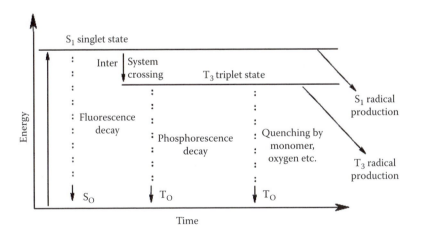

*Figure 1.8* Jablonski diagram. Energy levels vs. time.

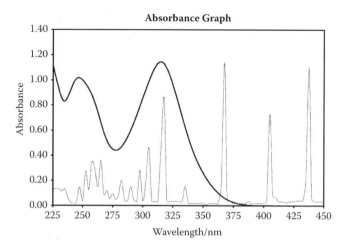

*Figure 1.9* Typical UV absorption curve of a photoinitiator (Speedcure BMS).

absorption must take place at all levels throughout the film layer for full depth cure to be effected.

The light absorption of a photoinitiator can conveniently be described by its absorption spectrum, where absorbance or optical density at a fixed concentration is recorded in a graph against wavelength. Figure 1.9 shows the absorption spectrum of 4-tolylthiobenzophenone (Speedcure BMS) against the background of the output lines of a medium-pressure mercury lamp.

A photoinitiator will display one or more absorption peaks at various wavelengths where the conversion of UV energy will be the most efficient. Absorption at short wavelengths around 250 nm is usually associated with $\pi \to \pi*$ excitation of the $\pi$ bonding orbitals, and long wave absorption around 330–420 nm comes from the n $\to$ $\pi*$ excitation of the n bonding orbitals.

The absorbance A, or optical density, is defined as the log of the incident light intensity $I_o$ over the transmitted light intensity $I_t$.

$$A = \log I_o/I_t$$

The absorbance is linearly related to the concentration of the solution (c) and the path length of the cell or depth of penetration through the film (d). This is the Beer–Lambert law.[3]

$$A = \varepsilon \ c \ d$$

If c is in moles per liter and d in cm, and A is measured at the wavelength of maximum absorption $\lambda_{max}$, then the molar absorption coefficient $\varepsilon_{max}$ of the photoinitiator can be calculated (liters/mol/cm).

The light intensity at depth d, and hence the radical count in a UV curable film, can be calculated from these equations.

To determine the depth to which light will penetrate to give full cure, these calculations should be carried out at all the wavelengths of the main output lines of the lamp that is used, since the absorption by the photoinitiator varies with wavelength. In practice, this calculation also needs to take in the absorption by the binders, fillers, pigments, and reflection and scattering effects, etc.

## 1.4   UV safety and ozone

UV radiation is non-ionizing at the wavelengths generally used for UV curing (above 200 nm) but should be regarded as hazardous, and appropriate safety equipment such as protective goggles should be worn.[17]

UV A lamps (315–400 nm) can be used to increase skin pigmentation and produce a tanning effect. UV B (280–315 nm) is more dangerous and can be erythrogenic, producing reddening of the skin and eyes, and lead to other effects such as blistering and conjunctivitis or snow blindness. Sunscreens, which include materials such as aminobenzoate esters, absorb UV in this region and will protect the skin from harm.

UV C (200–280 nm) can damage the living cell structure and has sterilizing properties. Short wave germicidal lamps are used in medicinal and food packaging applications. In daylight, UV C is normally filtered out by the ozone layer.

At very short wavelengths around 100 nm and below, the UV becomes sufficiently high in energy to take on some ionizing nature. It is essential to take every precaution and wear suitable eye protection while using any type of UV lamp.

Ozone is a toxic gas that can be produced by irradiation of oxygen (in the air) by short wave UV below 250 nm. UV lamps that do not produce these short wavelengths can be considered to be relatively safe with respect to the production of ozone, but most lamp units have efficient extraction from the unit to minimize any hazard from the buildup of ozone.

## References

1. Holman, R. A visit to the UV region. *PRA RADNews* 27 (Winter 1998): 21–23.
2. Caiger, N. and S. Herlihy. Energy curing in inkjet digital production printing. *SGIA Journal*, Fourth Quarter 2005, 35–39. www.sgia.org.
3. Skinner, D. (Fusion UV Systems Inc.). Efficient curing of performance coatings using high peak irradiance UV light. *Surf. Coat. Int. (JOCCA)* 83, No. 10, Oct. 2000, 508–511.
4. Zinnbauer, F. E. (Honle UV) UV curing and heat management for printing film substrates. *Proc. PRA Mat. Markets Conf.* Bradbury (UK), 2002. Paper 7.

5. Gingeri, F. (ELMAG S.p.A.) Cold cure system: A new generation of cold UV radiation sources. *RadTech Eur. Conf. Proc.*. 1991. Paper 43, 547–561.
6. Stowe, R. W. (Fusion UV) Dichroic reflectors applied to high peak irradiance microwave powered UV lamps. *RadTech NA. Conf. Proc.* 1992. 447–452.
7. Technical Data Sheets, Fusion UV Ltd. www.fusionuv.com/bulbs.
8. Skinner, D. (Fusion). Performance of photoinitiator systems with selected UV spectral output. *RadTech NA. Conf. Proc* 1994. 239–245.
9. Aldridge, A. D., P. D. Francis, J. Hutchinson. (Electricity Council Research Centre) UV curing of $TiO_2$ pigmented coatings—evaluation of lamps with differing spectral characteristics. *J. Oil Color Chem. Assoc.*, 1984 (2), 33–39.
10. Howell, B. F. and M. O'Donnell. Radiation curing of coatings with a xenon lamp. *RadTech NA. Conf. Proc.* 1994. 436–442.
11. Stropp, J. P., U. Wolff, W. Schlesing, S. Konaghan, H. Loffler, M. Osterhold, H. Thomas. (DuPont) UV curing systems for automotive refinish applications. *RadTech Eu. Conf. Proc.* 2005. 217–220.
12. Mills, P. (Phoseon) Characterising the curing capabilities of UV LEDs. *RadTech Eu. Conf. Proc.* 2005. 159–168.
13. Brandl, B. (IST Metz GmbH) UV–LEDs–Survey of a new emerging technology. *RadTech Eu. Conf. Proc.* 2007.
14. Decker, C. (ENSC Mulhouse) Recent advances in laser-induced curing. *RadTech Eu. Conf. Proc.* 1991. Paper 42, 536–546.
15. Nozaki, K. and E. Yano. High performance resist materials for ArF Excimer Laser and electron beam lithography. *Fujitsu Sci. Tech. J.*, 38, 1, p 3–12 (2002).
16. Hanrahan, M. J. (EM Industries). Oxygen inhibition. *RadTech Report.* March/April, 1990.
17. Schaper, K. L. (PPG Ind.) UV radiation: Effects on skin and eyes. *RadTech NA. Conf. Proc.* 1988. 34–36.

# chapter two

# A little chemistry

Two types of chemical processes are presently in general use for UV curing: free radical and cationic. Free radical chemistry, using photoinitiators to produce radicals from the energy of a UV light source, is by far the most common system for commercial applications where acrylate and methacrylate oligomers are formulated and polymerized.

Cationic curing (Chapter 7) involves the production of a Brønsted or Lewis acid from the photoinitiator via UV energy. The proton that is generated will open rings such as epoxys, oxetanes, etc., and initiate a ring opening polymerization process. The materials for this latter process are inherently more expensive than that of the acrylate chemistry used in free radical curing, and cationic curing is the minor partner in UV curable commercial applications, although it can bring enhanced properties for some applications such as metal decorating and plastics, due to the different types of polymers that are formed.

Other chemistries, such as the thiol-ene process (Section 2.8), are emerging but are presently limited to specific applications and not yet widely applied.

## 2.1 Free radical chemistry

This chapter examines the structure of the photoinitiator and the mechanism by which a reactive species is formed. This leads to the polymerization process involving acrylate-type materials. Substitution effects on the photoinitiator molecule are examined, with respect to UV absorption patterns and reactivity, giving an understanding of the wide variety of structures available.

Commercial photoinitiators in the text are referred to by their trade names as well as their chemical nomenclature and are followed by their numbers (n) in the tables.

The generation of free radicals from the absorption of UV light by the photoinitiator follows two distinct mechanisms, Type I and Type II, emanating from an excited species that is formed first.

## 2.2    The chromophore and absorption of UV energy

### 2.2.1    The generation of radicals

To produce free radicals, the photoinitiator has first to absorb energy from the light source. Most commercial photoinitiators have structures that absorb energy in the UV region between 200 nm and 400 nm, and the energy absorbed is sufficient to form an excited species. The chromophore that does this most efficiently is the aryl ketone group shown in Figure 2.1.

Substitution on the ketone at $R_1$ influences the mechanism by which free radicals are formed. A (substituted) alkyl group at $R_1$ will lead to a Type I scission process, whereas a (substituted) aryl group at $R_1$ leads to a Type II abstraction process. Substitution at $R_2$ will influence the wavelength at which light is absorbed by the photointiator in both types of mechanisms.

Absorption of UV energy by the chromophore initially produces an excited state in which the energy of the carbonyl group is raised to a higher level than that of the normal ground state of the molecule. An excited triplet state is formed (via a singlet).

Most photoinitiators produce free radicals from this excited triplet state in the process:

$$P \text{ (ground state)} + UV \rightarrow P^* \text{ (singlet)} \rightarrow P^* \text{ (triplet)} \rightarrow \text{Radicals}$$

When $R_1$ is an alkyl group, the CO-alkyl bond energy is of the order of 65–70 kcals/mol. UV light can provide energies of around 70–80 kcal/mol, which is sufficient to cleave the CO-alkyl bond and produce two free radicals.

$$P^*\text{(triplet state)} \rightarrow \text{Benzoyl radical} + \text{Alkyl radical}$$

This occurs, for example, with Darocur 1173 (1). Figure 2.2 illustrates the Type I scission process.[1]

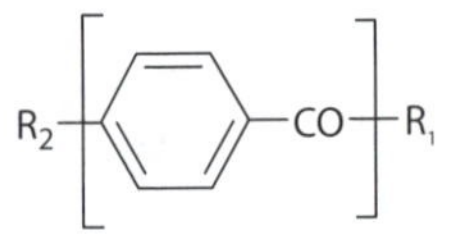

*Figure 2.1* The aryl ketone group.

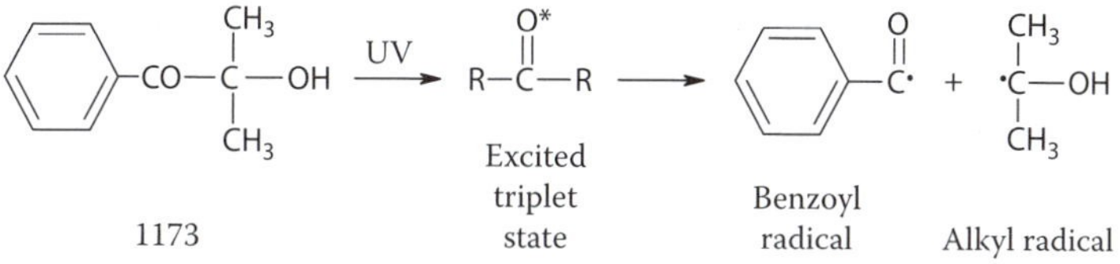

*Figure 2.2* The Type I unimolecular scission process (Norrish Type I).[1]

When $R_1$ is an aryl group, the CO-aryl bond energy is a little higher at 80–90 kcal/mol and the available UV energy is insufficient to split this bond. In this case, the photoinitiator absorbs UV energy as before, but remains in an excited triplet state until it can approach and react with a suitable hydrogen donor. Hydrogen donors are molecules such as tertiary amines, ethers, esters, thiols, etc., that have activated hydrogen atoms alpha to a hetero atom.

The triplet state is able to react with a hydrogen donor, taking a hydrogen atom to produce a ketyl radical, which has a low reactivity, and a donor radical that is highly reactive.

$$P^* \text{ (triplet state)} + HD \text{ (donor)} \rightarrow P^{\cdot} H \text{ (ketyl radical)} + D^{\cdot} \text{ (donor radical)}$$

Figure 2.3 illustrates the Type II hydrogen donation process that typically occurs, for example, with benzophenone (50) and its derivatives.[3]

The excited triplet state of the initiator is a transient state. It can go on to produce radicals via both mechanisms, as described previously, or it can decay back to the ground state with loss of energy in the form of phosphorescence. Deactivating processes can also occur, such as monomer and oxygen quenching, etc., which compete with the formation of radicals. For example, the triplet state of the initiator can be reduced to the ground state

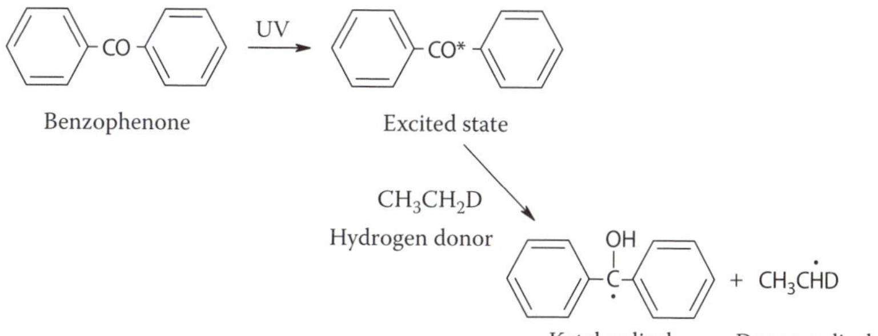

*Figure 2.3* The Type II bimolecular hydrogen abstraction process.[2]

by oxygen in the film and the subsequent excited singlet oxygen state then simply relaxes back to the stable triplet form of oxygen. The overall effect is simply a loss of energy.

$$P^* + O_2 \rightarrow P + O_2^* \rightarrow O_2 + heat$$

Triplet lifetimes are very short for Type I photoinitiators prior to scission, so the competing quenching processes are less likely to be of influence. Monomers such as styrene and unsaturated polyesters are strong triplet quenchers and Type I photoinitiators with very short triplet lifetimes are most efficient for formulations containing these materials. Type II photoinitiators have longer triplet lifetimes, which allows time for bimolecular hydrogen donor reactions to take place, but also gives time for quenching reactions. Type II photoinitiators are therefore not suitable for use with styrene or formulations containing strong triplet quenchers.

## 2.2.2   The polymerization process

The generation of free radicals outlined previously is the first step in the UV curing process. Monomers and oligomers are then required to produce a polymeric matrix from the formulation to be cured. Free radical photopolymerization involves four processes: initiation, propagation, chain transfer, and termination. Only the initiation step is photosensitive; all the other steps are thermally driven.

1. Initiation comprises the process of absorption of UV energy by the photoinitiator, the formation of radicals, and the reaction of a radical with an acrylate species to create a reactive alkyl monomer radical.

$$I \text{ (initiator)} + UV \rightarrow R^\cdot \text{ (radical)}$$

$$R^\cdot + CH_2 = CHCOOR \rightarrow RCH_2C^\cdot HCOOR$$

(acrylate monomer)       (monomer radical RM$^\cdot$)

The active radical produced by the photoinitiator becomes part of the reactive monomer and is locked into the growing polymer chain. The first initiation step, generating radicals, will stop when the light is switched off.

2. Propagation involves the monomer radical starting a chain reaction with the monomer and oligomer in the formulation. Cross-linking occurs if multifunctional monomers are used.

$$RCH_2\overset{\cdot}{C}HCOOR \; + \; CH_2{=}CHCOOR \longrightarrow RCH_2CH{-}[CH_2CH]_n - CH_2\overset{\cdot}{C}H \text{ etc.}$$

monomer radical RM ·    acrylate monomer M
                        or oligomer O

$$\underset{\text{COOR}}{|} \quad \underset{\text{COOR}}{|} \quad \underset{\text{COOR}}{|}$$

$$RM[M]_nM\cdot$$

3. Chain transfer involves the reaction of the growing polymer radical with a hydrogen donor that transfers a hydrogen atom to the polymer chain, stopping its growth, and creates a new donor radical that can generate a new polymer chain by reacting with more monomer. The use of hydrogen donors, as well as being an essential part of the Type II initiation process, can modify and reduce the average molecular weight of the polymer that is formed by forming multiple polymeric species by chain transfer.

$$RM_n\cdot + DH \rightarrow RM_nH + D\cdot \quad D\cdot + M \rightarrow DM\cdot \text{ etc.}$$

4. Termination occurs where radicals interact or recombine to give a neutral species and the polymer chain stops growing.

$$RMM\cdot + RMM\cdot \rightarrow RMM{-}MMR \quad RMM\cdot + R\cdot \rightarrow RMM{-}R \text{ etc.}$$

Radicals from various sources may interact. The ketyl radical from a Type II initiator is often a cause of chain termination. Polymerization stops of necessity when all the monomer is used, but more usually occurs when the increasing viscosity of the coating precludes further molecular movement and interaction, despite the fact that there may still be some unsaturation left in the system.

The efficiency and the speed of the polymerization process will depend on many factors and is related to the formulation as a whole, where binders, monomers, pigments, additives, etc., all influence the curing process. In addition, the viscosity of the formulation can have a profound effect where very low viscosities, such as in UV flexo, allow oxygen to diffuse back into the film at a high rate and retard the curing process. UV inkjet is even more difficult to cure; whereas the highly viscous, buttery nature of UV offset allows relatively easy cure.

Some photoinitiators, such as the alkylaminoacetophenones (AAAPs), are much more efficient in producing radicals than others and lead to very fast curing, but the relative cure speeds of photoinitiators will vary with the formulation. The functionality of the monomer also has a strong influence. Multi-functional monomers will react much faster than mono-functional monomers due to increased levels of cross-linking, and very hard coatings can be formed. Other factors that influence cure speed include light intensity, film thickness, the use of pigments and light scattering, the use of various hydrogen donors, substrate reflection, etc.

## 2.3   Type I photoinitiators: Mechanism of the scission process

Looking at the scission process in more detail, scission can occur at any weak bond, not necessarily connected to the carbonyl function, and will generally occur at the weakest bond in the structure. In some cases, this may be a C–N bond or C–S bond and beta-scission occurs. In most cases, scission occurs alpha to the carbonyl group at the CO–C bond.

### 2.3.1   Alpha-scission

Scission at the α-carbon of the alkyl ketone, also called the Norrish Type I reaction, is by far the most common Type I mechanism.

A (substituted) alkyl group at $R_1$ will lead to a scission process.

UV light is absorbed by the photoinitiator to form an excited triplet state that goes on to form two active free radicals via cleavage of the CO–alkyl bond. For maximum efficiency, the alkyl group needs to be fully substituted (Figure 2.4).

The nature of the aryl substitution at $R_2$ and the alkyl substitution at $R_3$ will influence the wavelength at which light is absorbed and the efficiency of the cleavage process, respectively. Electron donors such as sulphur or nitrogen at $R_2$ will red shift the absorption to longer wavelengths. A hetero atom such as O or N on the alkyl group at $R_3$ will greatly increase the rate of cleavage.

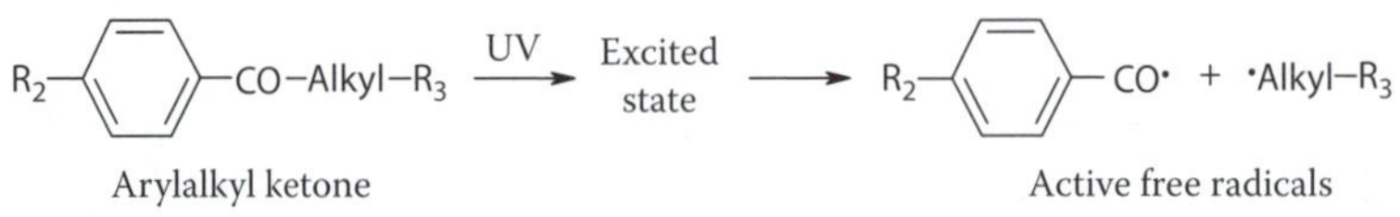

$$R_2 - \langle \rangle - CO - Alkyl - R_3 \xrightarrow{UV} \text{Excited state} \longrightarrow R_2 - \langle \rangle - CO^{\bullet} + {}^{\bullet}Alkyl - R_3$$

Arylalkyl ketone                                                        Active free radicals

*Figure 2.4* The Type I α-scission process.

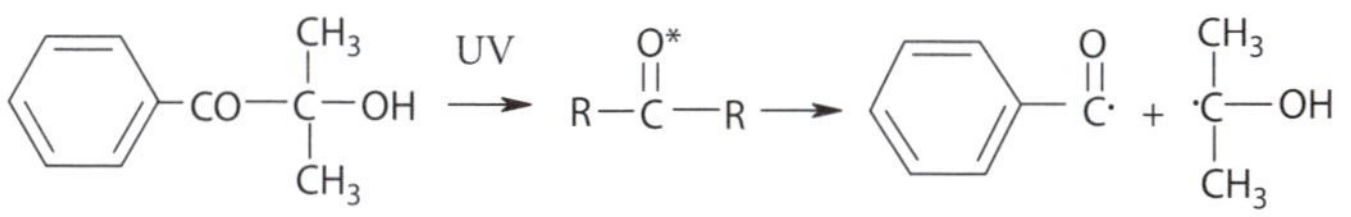


1173

Substituted acetophenone

Excited triplet state

Benzoyl radical

Alkyl radical

Both highly reactive

***Figure 2.5*** The formation of free radicals from 2,2-dimethyl-2-hydroxyacetophenone (1).

The scission process for the substituted acetophenone photoinitiator 2,2-dimethyl-2-hydroxyacetophenone, Darocur 1173, (1) is well documented (Figure 2.5). Both the benzoyl and the alkyl radical are reactive species.

## 2.3.2 *Beta-scission*

β-scission occurs when there is a weak bond from the alpha carbon to a hetero atom, which may be C–Cl, C–S or C–N, etc. (Figure 2.6).

A weak bond in the β-position, such as C-Cl, makes it susceptible to cleavage and scission will occur at this point. Both the dichloromethyl radical (α-carbon) and the chlorine radical are reactive. The chlorine radical may abstract a hydrogen atom from a donor and produce hydrogen chloride in the cured formulation. The formation of acid in this way may be detrimental to the application. Similar photoinitiators showing β-scission include beta-sulphonyl ketones (62) and trichloromethyl-S-triazines (27).

The alkylaminoacetophenone photoinitiator Irgacure 907 (10) cleaves mainly by alpha-scission but some degree of beta-scission of the weaker C–N bond also occurs,[4] producing different radicals (Figure 2.7).

Alpha scission produces the normal (substituted) benzoyl radical plus an alkylamino radical, both of which are very reactive. A small amount of beta-scission produces a phenacyl radical and a morpholino radical with different properties, though both are reactive.

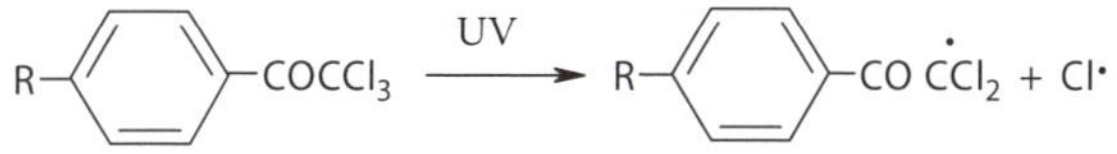


α-Trichloromethyl ketone

α-Carbon radical

Chlorine radical

***Figure 2.6*** Beta-scission of a trichloromethylacetophenone (25).

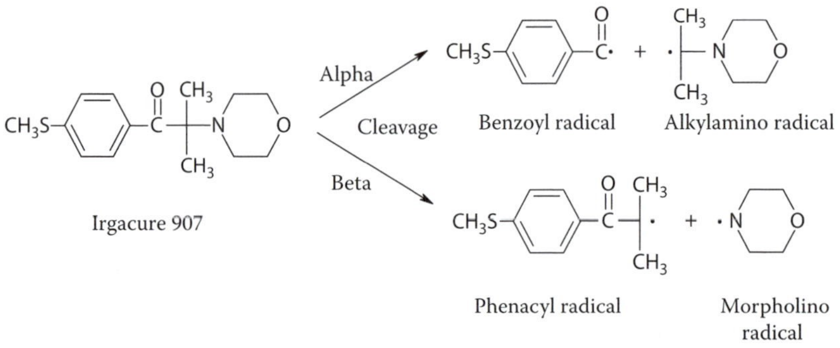

*Figure 2.7* Alpha and beta cleavage of Irgacure 907.

Irgacure 369 (12) and 379 (14) also show a small amount of beta cleavage of the C–N bond similar to the above, producing a dimethylamino radical, which is more reactive than the morpholino radical.

## 2.4   Type II photoinitiators: Mechanism of the abstraction process

Benzophenone has long been known as a triplet energy source that interacts with a hydrogen donor to produce radicals. It has been studied widely and has been used since the very early days of UV curing.

### 2.4.1   Hydrogen abstraction from a donor molecule

The excited triplet states of aryl–aryl ketones such as benzophenone (50), where $R_1$ is a (substituted) aryl group, possess CO–aryl bond energies that are too high for the available UV energy to break and a hydrogen abstraction process takes place from the long lived n–π* excited triplet state.

A donor molecule is needed to produce free radicals. Donor molecules usually contain hetero-atoms with active hydrogen atoms in the α-position. Tertiary amines, alcohols, ethers, esters, thiols, etc. are often used as donors. The donor transfers a hydrogen atom to the excited photoinitiator and becomes a very active donor radical.

Direct hydrogen abstraction (Figure 2.8) takes place from donors such as ethers or alcohols. Tertiary amines react more efficiently via an electron transfer mechanism that initially occurs from the donor nitrogen atom, and a transient excited state, or exciplex, is formed. Proton transfer follows from the alpha carbon atom on the donor to produce two free radicals (Figure 2.9). In this type of mechanism the π–π* triplet state is equally as effective as the n–π* state.

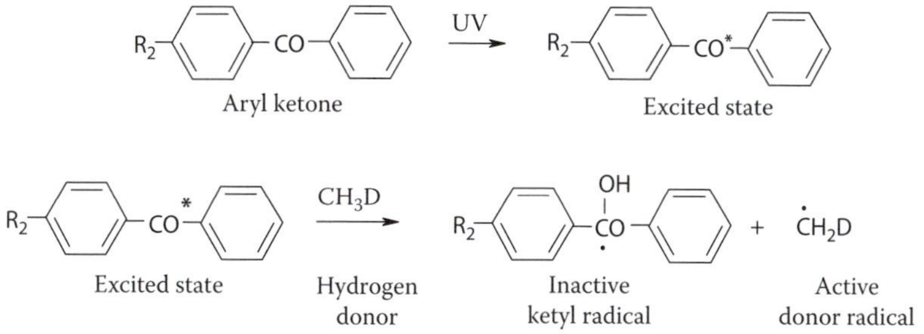

**Figure 2.8** The Type II hydrogen abstraction process.

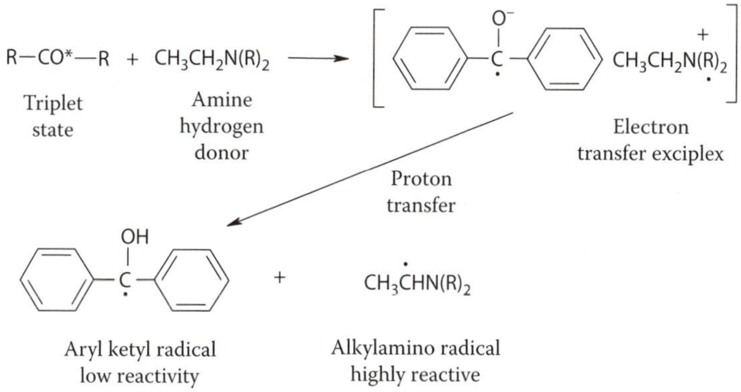

**Figure 2.9** The reaction of benzophenone triplet state with a tertiary amine.

The aryl ketyl radical has a low reactivity toward acrylate/methacrylate polymerization and can be a source of termination, limiting the polymerization process. The alkylamino radicals from the tertiary amine are highly reactive and will start the polymerization, becoming part of the growing polymer chain. Hence, the structure of the tertiary amine, which forms the active radical, has significant influence on the reactivity of a Type II system.

## 2.4.2 *Intramolecular γ-hydrogen abstraction: The Norrish Type II reaction*

2,2-Diethoxyacetopheneone, DEAP (18), is one of a few photoinitiators that will undergo a Norrish Type II intramolecular cyclization process,[5] as illustrated in Figure 2.10, in competition with the Type I scission process.

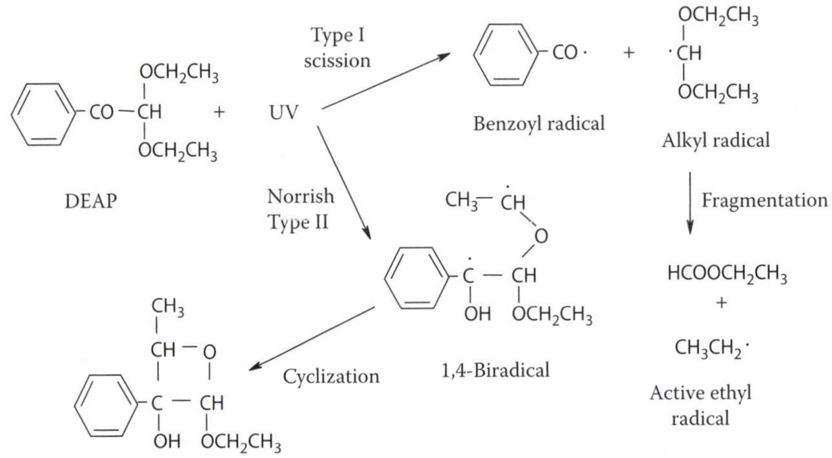

*Figure 2.10* The Norrish Type II intramolecular hydrogen abstraction process.

DEAP can undergo the standard Type I scission to produce two active radicals. However, the excited triplet state of photoinitiators that contain an activated $\gamma$-hydrogen atom, such as the ether structure in DEAP, also produces a 1,4-biradical by hydrogen abstraction from the $\gamma$–hydrogen on the ether. The biradical is not very efficient as an initiator of polymerization and tends to cyclize to a non-productive molecule. This competitive reaction makes DEAP a less efficient photoinitiator. The alkyl ether radical that is initially produced via Type I scission fragments thermally to give a reactive ethyl radical and ethyl formate.

Methyl benzoylformate, MBF (73), also undergoes a Norrish Type II intramolecular hydrogen abstraction process to produce a biradical (see Section 2.7).

## 2.5   The influence of molecular substitution on absorption and photoactivity

The photoactivity of a formulation is related to many factors including the UV dose, the wavelength and output of the UV lamps, the absorption characteristics of the photoinitiator, and the reactivity or unsaturation of the resin environment. Here, we are looking at substitution patterns on the initiator that will influence the efficiency of radical production in a standard environment. By modifying these substitutions, a large number of photoinitiators possessing a variety of UV absorption patterns and varied photoactivities have been made.

## 2.5.1 Type I photoinitiators: Substitution effects

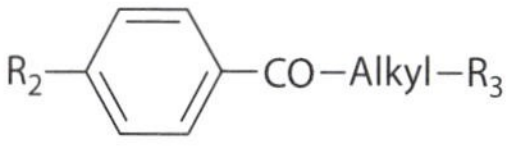

Alkyl aryl ketone

For Type I photoinitiators, electron donating substitution on the aryl ring at $R_2$ will red shift the absorption to longer wavelengths. Electron donating substitution on the alpha carbon at $R_3$ increases the rate of scission and generally leads to higher photoactivity.

For example, with hydroxyacetophenones, HAPs, such as Darocur 1173 (1), shown in Figure 2.11, alternative oxygen substitution on the alpha carbon (at $R_3$) also has some influence on reactivity. An ether group gives similar reactivity to the hydroxy group but an ester group has an electron withdrawing effect and the reactivity of the photoinitiator is much reduced. Nitrogen substitution instead of oxygen at $R_3$ leads to much improved photoactivity, as shown by the alkylaminoacetophenones (AAAPs).

Table 2.1 shows the effect of stronger electron donation on the aryl group at $R_2$ leading to a red shift in absorption from 248 nm (unsubstituted) to 325 nm with the morpholino substitution. Alternatively, para-dimethylamino substitution (at $R_2$) leads to a much longer triplet lifetime, allowing deactivating quenching processes to occur, and produces a much less efficient photoinitiator.

Photoactivity with these particular products increases along with the absorption wavelength and follows increasing electron donating substitution on the alpha carbon at $R_3$, although there are many other factors that

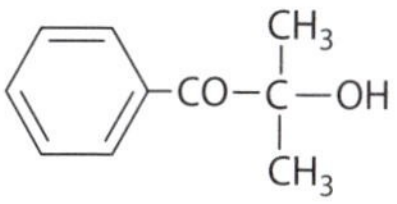

*Figure 2.11* Darocur 1173.

*Table 2.1* Type I Substitution and Wavelength Shift

| Substitution | | Trade Name | Absorption (nm) |
|---|---|---|---|
| $R_2$ | $R_3$ | | |
| H | OH | Darocur 1173 | 248 |
| CH$_3$S | 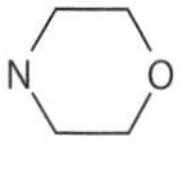 N⌿O | Irgacure 907 | 306 |
| 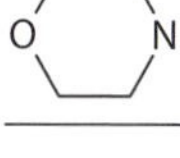 O⌿N | (CH$_3$)$_2$N | Irgacure 369 | 325 |

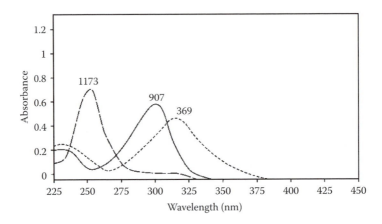

*Figure 2.12* UV absorption of 1173, 907, and 369.

influence reactivity, and longer wavelength absorption is not always the prime factor.

Photoactivity increases in the order: 1173 <907 <369.

Figure 2.12 shows the effect on absorption patterns of substitution on the aryl group at $R_2$ on the hydroxyl or alkylamino substituted acetophenone type of structure. The UV absorption is red shifted from 248 nm (Darocur 1173) to 306 nm (Irgacure 907) and 325 nm (Irgacure 369) with increasing electron donation at $R_2$.

At the same time, the efficiency of the scission process is influenced by electron donation on the alpha carbon at $R_3$ and the photoactivity and cure speeds obtainable from these photoinitiators increases. The long tail of Irgacure 369 toward 400 nm is the result of bulkier alkyl substitution on the alpha carbon atom (2-benzyl-2-ethyl vs. 2,2-dimethyl for Irgacure 907).

Chapter 3 gives a more detailed account of substitution effects and the design of photoinitiators.

### 2.5.2   Type II photoinitiators: Substitution effects

$$R_2 \text{—} \langle\text{aryl}\rangle \text{—CO—} \langle\text{aryl}\rangle \text{—} R_3$$

Aryl ketone

For Type II photoinitiators, electron donating substitution on either aryl group at $R_2$ or $R_3$ will produce a similar red shift of wavelength and activating effect. Electron withdrawing substitution has a negative effect on the photoactivity and, for example, derivatives based on benzophenone tetracarboxylic dianhydride are less photoactive.

*Table 2.2* Type II Substitution and Wavelength Shift

| Substitution | | Trade Name | Absorption (nm) |
|---|---|---|---|
| R₂ | R₃ | | |
| H | H | Benzophenone | 254 |
| H | ⬡ | Speedcure PBZ | 288 |
| H | S—⬡—CH₃ | Speedcure BMS | 315 |

With benzophenones, Table 2.2 shows the influence of increasing electron donating substitution on wavelength.

Photoactivity increases in the order: BP <PBZ <BMS.

Figure 2.13 shows absorption shifts similar to those of Type I, although the reactivity difference between PBZ and BMS is not great.

The absorption figures quoted are the wavelengths of maximum absorption. UV absorption tends to follow a bell-shaped curve that gives an absorption band over which light energy can be absorbed. Measurement of absorption maxima is also influenced to a small degree by the polarity of the solvent that is used in the process, and leads to a small range of quoted figures for wavelength absorption vs. solvent.

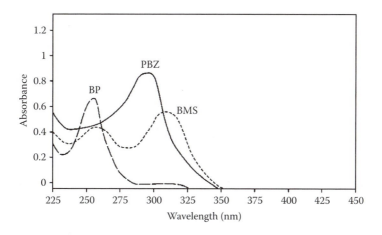

*Figure 2.13* UV absorption of BP, PBZ, and BMS.

### 2.5.3   Commercial photoinitiators

The wide variety of commercially available photoinitiators gives a large range of absorption profiles and these will cover in total the whole spectrum of UV and near-visible light with the premise of allowing

- Matching of the initiator absorption to the wavelength of the light source
- Variety of performance criteria such as cure speed, surface cure, depth cure, and yellowing
- Formulations to serve a multitude of applications
- Commercial economy

For maximum efficiency, it is essential that the photoinitiator can pick up one of the stronger output lines that the lamp in use will provide. Matching the photoinitiator to the lamp output should always be a prime consideration. The MPM lamp provides energy over most of the UV spectrum and is the UV lamp most commonly in use. The different types of commercial photoinitiators can be related to the output of a MPM lamp, as illustrated in Figure 2.14.

The hydroxyacetophenones, benzil ketals, alkylaminoacetophenones and phosphine oxides will provide absorption throughout the UV spectrum using Type I photoinitiators (Figure 2.14).

The benzophenones, substituted benzophenones, anthraquinones and thioxanthones will provide a similar range of UV absorption using Type II photoinitiators (Figure 2.15).

In general, photoinitiators that absorb in the short wave UV around 250 nm are used in clear coatings and varnishes that require low yellowing properties. Photoinitiators that absorb in the long wave UV at 300–400 nm have greater use in ink systems where pigment absorption has to be taken into account. Photoinitiators that absorb above 400 nm, in the near-visible range, will absorb the blue end of the visible spectrum and take on a yellow coloration. Hence, all thioxanthones, which absorb

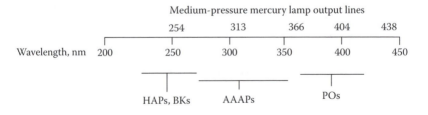

*Figure 2.14* Type I photoinitiators vs. wavelength.

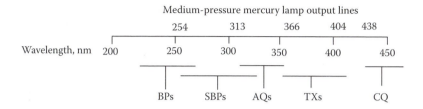

*Figure 2.15* Type II photoinitiators vs. wavelength.

to a small degree above 400 nm, are yellow and will impart some yellowness to the formulation and to the cured product. Although the phosphine oxides also absorb in the near-visible range and are yellow materials, their photobleaching properties mean that they lose their color under exposure to UV or visible light.

## 2.6 Photobleaching

Photoinitiators that absorb above 400 nm absorb the blue end of the visible spectrum and are yellow in color. Some photoinitiators, such as the Type I phosphine oxides, cleave to produce two radicals, neither of which absorb UV above 400 nm. This means that, as light is absorbed, the chromophore is destroyed, as shown in Figure 2.16, and colorless photoproducts are produced. The original yellow color of the phosphine oxide disappears with photolysis, becoming colorless.[6]

This photobleaching effect has two advantages. On the one hand, as long wave UV penetrates the film, the coating becomes transparent, and light is able to penetrate farther into the coating to allow cure to a considerable depth. Second, continuous exposure to UV light around 400 nm, or to visible light, will bleach the cured film to give a colorless product.

Photoinitiators that photobleach provide excellent depth cure.

The phosphine oxides, TPO (22), TPO-L (23) and 819 (24), photobleach and give excellent depth cure and colorless coatings.

Figure 2.17 shows the scission of TPO producing two radicals. The phosphinyl radical is very reactive but suffers from being very susceptible to oxygen quenching, which makes the phosphine oxides unsuitable for thin film curing.

Although long wave photobleaching by the phosphine oxides is by far the most common and useful effect, photobleaching is not necessarily confined to long wave UV. 1,2-Diones such as benzil will photobleach as the conjugated carbonyl structure is destroyed on photolysis, allowing further light penetration. Benzil (76), absorbing at 260 nm, is however, a relatively poor photoinitiator and is seldom used.

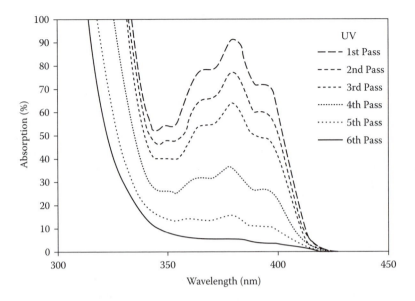

*Figure 2.16* Photobleaching of the long wavelength absorption of TPO. (Adapted from W. Reich et al. *RadTech NA 2000 Conference Proceedings*.)

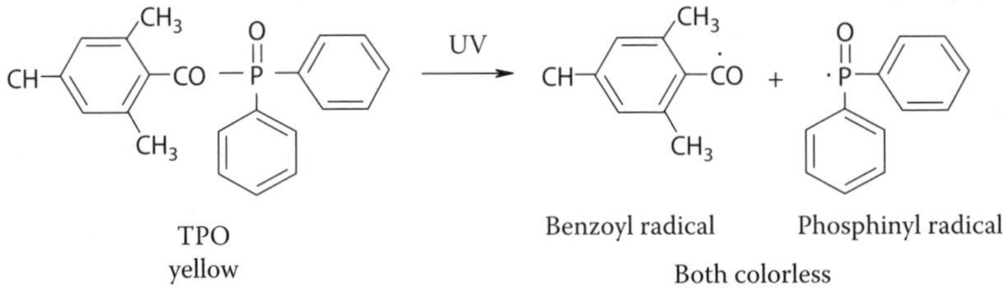

*Figure 2.17* α-scission of phosphine oxides.

Camphorquinone, CQ (75), a 1,2-dione that absorbs in the visible spectrum at 468 nm, is very yellow. Photolysis destroys the 1,2-conjugation with its associated absorption and effectively causes photobleaching. CQ is a relatively slow-curing photoinitiator and its major use is in dental applications, where the photobleaching leads to excellent whites and cure speed is not a major factor.

Thioxanthones (63–72), which also absorb UV above 400 nm, are Type II initiators that do not cleave. Unless there is a hydrogen donor available to effect a Type II reaction, absorption of UV by the thioxanthone in the cured film will have no effect on its structure, and the original yellow from any unreacted thioxanthone will remain. Thioxanthones therefore do not photobleach.

## 2.7 Diverse mechanisms: Variations on the Type I and Type II process

The majority of commercial photoinitiators follow either a Type I or a Type II mechanism but there are a few photoinitiators whose mechanism of radical production follows variations of these.

### 2.7.1 Acyloximino esters

Acyloximino esters undergo a Type I scission of the N-O bond in the gamma position.[7]

Figure 2.18 shows the fast fragmentation of the N-O bond (γ-scission) to form an oximino radical, which is the prime initiator of polymerization. This Type I scission process occurs at the weakest point in the molecule, the N-O bond, making it a gamma cleavage reaction. Further fragmentation of the radicals that are first produced leads to benzoyl and ethoxy radicals, both of which are reactive, and produces traces of acetonitrile and carbon dioxide as photoproducts. Speedcure PDO (35), being a Type I photoinitiator, does not require a hydrogen donor. Irgacure OXE photoinitiators (36 & 37) are oximino photoinitiators similar to PDO, with substitution patterns designed specifically to absorb at longer wavelengths for the production of color filter matrices.

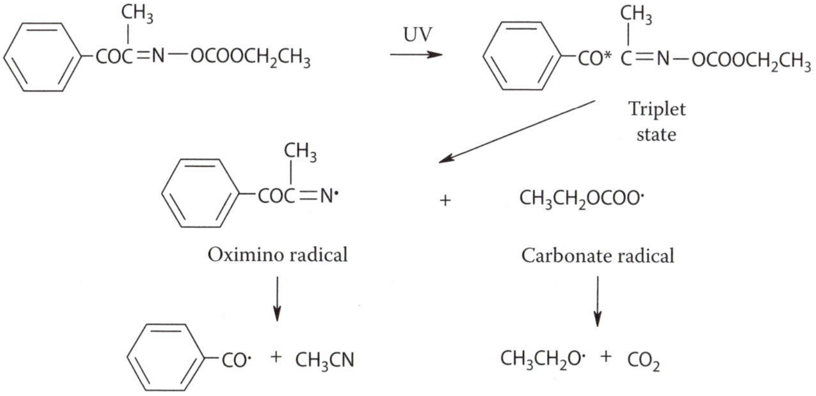

*Figure 2.18* Radical formation from 1-phenyl-1,2-propanedione-2-(O-ethoxycarbonyl)oxime (PDO).

## 2.7.2  Anthraquinones

Anthraquinones follow the Type II hydrogen abstraction process but are more efficient in the presence of oligomeric hydrogen donors than with tertiary amines.[8]

2-Ethylanthraquinone, EAQ (77), is a Type II photoinitiator that requires a hydrogen donor. However, EAQ is less effective when used with a tertiary amine than with ethers, esters, etc., or activated hydrogen atoms within the resin matrix. The EAQ abstracts a hydrogen atom to produce a donor radical that is the main reactive species, as shown in Figure 2.19. The EAQ becomes a semi-reduced quinone that disproportionates to produce EAQ and 2-ethyl-5,9-dihydroxyanthracene. The latter is readily oxidized back to EAQ by any oxygen that is in the system. Oxidation happens during cure and depletes the oxygen in the coating, regenerating EAQ and making this photoinitiator less sensitive to oxygen inhibition.

EAQ is used widely in the electronics sector for acid resists, where tertiary amines cannot be used and where the oligomeric backbone of urethane and epoxyacrylates can be effective hydrogen donors. For surface coatings and for inks, EAQ has generally been replaced by the more efficient thioxanthones.

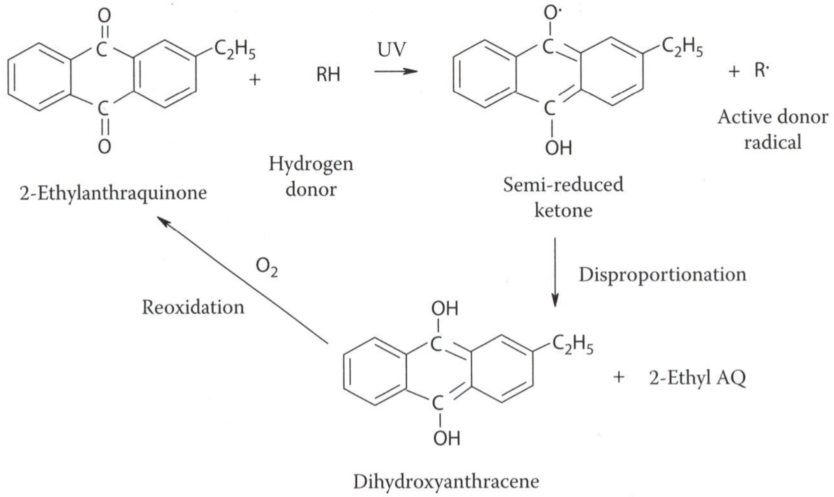

*Figure 2.19* Radical formation from 2-ethylanthraquinone.

## 2.7.3 BCIM and the lophyl radical

2,2-Bis(2-chlorophenyl)-4,4',5,5'-tetraphenyl-2'H-(1,2')-biimidazolyl (BCIM) (38) is one of a series of hexaaryl-bisimidazolyls or HABIs.

BCIM is an imidazole dimer that will cleave at the C-N bond under UV light to give two imidazole nitrogen radicals, or lophyl radicals, which are relatively stable and inactive[9] (Figure 2.20).

The lophyl radical undergoes electron transfer with leuco dyes to produce color by disproportionation of the leuco dye radical, and the system is widely used for direct-write laser imaging.

Alternatively, in a conventional photoinitiation process similar to a Type II, the lophyl radical will undergo radical exchange with a strong electron donor such as a thiol to produce an active thiyl radical that will initiate the polymerization of acrylate formulations.

HABIs can be used with thiols such as 2-mercaptobenzthiazole (133) or 2-mercapto-benzimidazole (134) as electron donors to provide the coinitiator in a standard Type II photoinitiator system. Tertiary amines such as the aminobenzoates (115–124) are less effective as coinitiators. If leuco crystal violet (LCV) (129) is then added to this system, it becomes a more efficient combination. Structural modifications such as electron withdrawing substituents on the aryl groups will improve the photoactivity of the HABIs and the o-chloro substitution confers ring twisting that increases electron acceptance of the lophyl radical and improves performance.

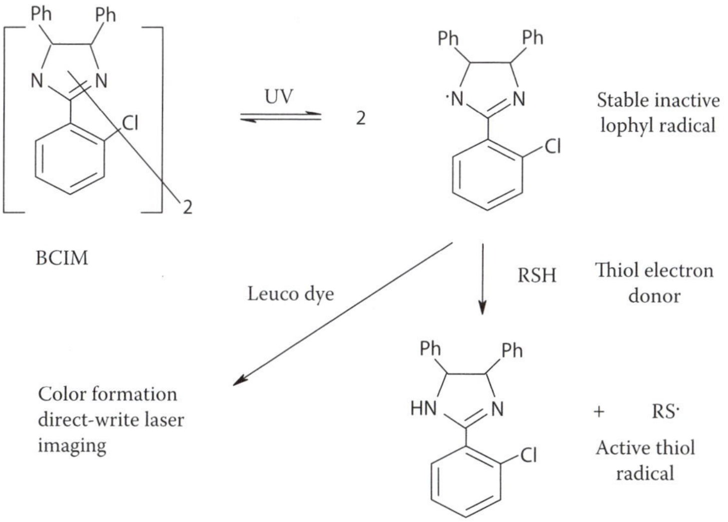

*Figure 2.20* The photolysis of HABIs and the lophyl radical.

## 2.7.4　Benzoylformate esters

Benzoyl formate esters follow the Norrish Type II hydrogen abstraction mechanism from the alkyl ester group, producing a biradical,[10] followed by an efficient radical transfer to a suitable oligomer donor (Figure 2.21).

Unlike the Type I alkylaryl ketones, scission does not occur to any great extent with methyl benzoylformate, although photoproducts relating to the benzoyl radical have been found. The dominant process is an intramolecular hydrogen abstraction process (Norrish Type II) taking place from the γ-hydrogen atom on the ester function, producing a 1,4-biradical that is only weakly reactive toward acrylates. The biradical can fragment or cyclize to give inert species, which is the usual reaction of a Norrish Type II, but it can also undergo rapid radical transfer with a hydrogen donor on the oligomer, such as an oligoether, to produce a reactive oligoether radical plus the normal inert ketyl radical of the Type II reaction. The oligo-radical can chain extend or transfer to give high cross-link density polymers.

The biradical is also an efficient oxygen scavenger and reduces the effects of oxygen inhibition.

Traces of photoproducts produced from the minor decomposition of the biradical may include benzaldehyde, formaldehyde, and carbon monoxide.

Methyl benzoylformate, MBF (73), is one of the few Type II photoinitiators that are less efficient with a tertiary amine as hydrogen donor. With formulations containing MBF, the presence of amines may lead to storage problems and the development of odor. MBF works very well in urethane acrylate oligomers to produce very hard, clear coatings. The bis(benzoylformate) ester Irgacure 754 (74) will follow a similar mechanism.

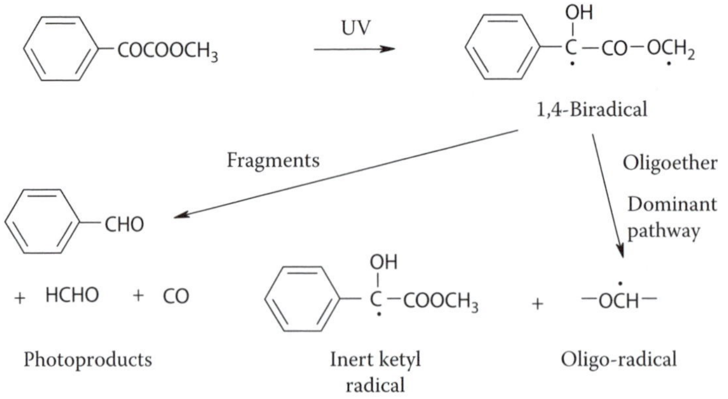

*Figure 2.21* Radical formation from methyl benzoylformate.

## 2.7.5   Substituted maleimides as photoinitiators

Substituted maleimides can behave as photoinitiators, either as a Type II mechanism in the presence of a hydrogen donor or in a Norrish Type II intramolecular donation giving a γ-biradical,[11] shown in Figure 2.22.

The efficiency of the reaction of N-alkyl maleimides with tertiary amines depends on the alkyl substituent, with cyclohexyl > ethyl > methyl > tert. butyl, the order of reactivity. N-aryl-substituted maleimides are much less reactive. The hydrogen abstraction process from the amine can result in a standard ketyl type radical being formed from the carbonyl group, or the unsaturation can be involved, resulting in an ene-derived radical. There is some evidence to suggest that the latter is a more significant process.[12] In the absence of a hydrogen donor, maleimides with gamma-hydrogen substituents can undergo an intramolecular reaction producing the 1,4-biradical, which has been shown to initiate acrylate polymerization but is less efficient than the intermolecular process with an amine.

N-substitution with a methylene spacer on an ester or carbonate group brings higher reactivity, approaching that of Irgacure 651, but a N-methyl urethane is less reactive.

The maleimides absorb at 295 nm and photobleach as the unsaturation reacts. They have been described as photoactive monomers in a "photoinitiator free" system. Interestingly, the maleimides can be sensitized by

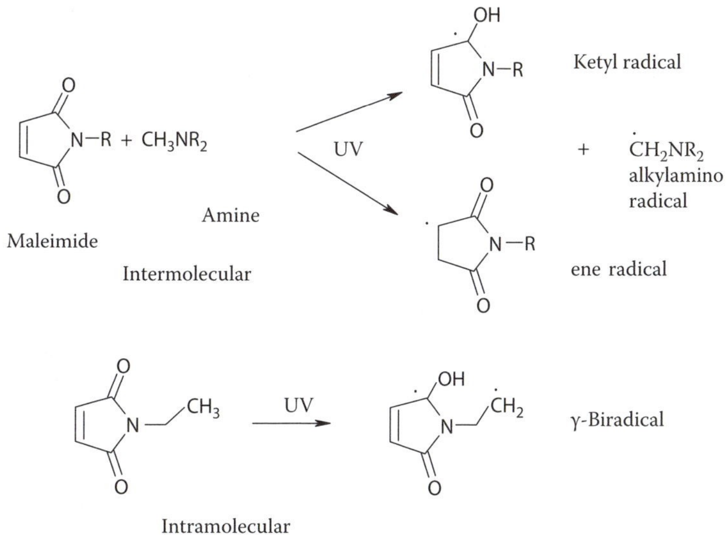

*Figure 2.22* Radical processes of N-substituted maleimides.

benzophenone or thioxanthone. The addition of a small amount of thioxanthone to a 1:2 mixture of N-methyl maleimide/MDEA gave triple the photo DSC peak exotherm.[13] In a similar experiment, the addition of 0.1% 2,3-dimethyl maleic anhydride to a 1:1 benzophenone/MDEA combination gave similar enhanced rates of polymerization.[14] Sterically hindered anhydrides are required to give the formulation sufficient stability.

No initiators based on the maleimide structure have been commercialized as yet,[15] possibly due to the toxicity of these materials, but mono- and difunctional maleimide monomers have been produced for clear coatings.

## 2.7.6  *Phosphine oxides and secondary scission*

Phosphine oxides such as Lucirin TPO (22) follow the standard Type I scission process to produce two radicals (Figure 2.17). Both the benzoyl radical and the phosphinyl radical are very reactive and will initiate the acrylate polymerization.

Bis-phosphine oxides such as Irgacure 819 (24) in theory can produce four radicals. A trimethylbenzoyl (1) radical and a trimethylbenzoylphosphinyl (2) radical are formed from the primary scission, both of which are active. The latter radical can undergo further scission, producing a second trimethylbenzoyl (3) radical and a (monomer)phosphinyl radical (4), making them very fast, efficient photoinitiators. In practice, secondary scission of the growing end-capped phosphinoyl oligomer is more likely to occur, as shown in Figure 2.23.[16]

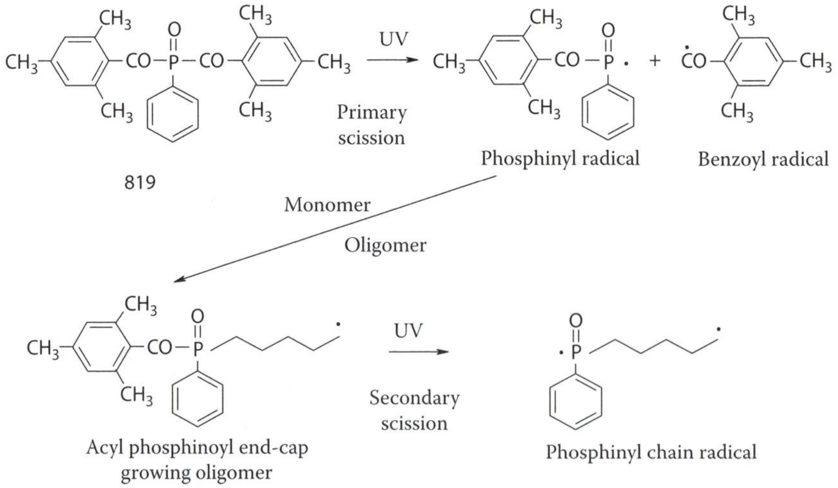

*Figure 2.23* Secondary α-scission of bis-phosphine oxides.

The trimethylbenzoylphosphinyl radical that is initially produced is very reactive and will start the chain reaction with the monomer and oligomer to produce an end-capped phosphinoyl group on the growing chain. This group has the ability to cleave under further irradiation and produce a second benzoyl radical plus a phosphinyl radical on the polymer that can extend the growing chain or lead to cross linking. The result is a longer chain, denser, polymeric structure.

## 2.7.7   Photo-acid generation

The generation of acid species via UV light has a variety of applications in the industry.[17]

The following schemes illustrate the types of molecules that can be employed to produce acids via a Type I photocleavage.

Arylketosulphinates and o-nitrobenzyl esters will cleave to produce sulphinic or sulphonic acids.

Arylketosulphinate photoinitiators, such as Esacure 1001M (62), under extended irradiation, can cleave at the beta position to release arylsulphinic radicals (Figure 2.24). These radicals are inefficient catalysts for acrylate polymerization, but readily undergo hydrogen abstraction in the presence of neutral donors such as ethers, esters, etc., to generate sulphinic acids. Similarly, the o-nitrobenzyl esters can produce p-toluenesulphonic acid, as shown in Figure 2.25.

The sulphonic acids usually require a thermal step to complete the polymerization of epoxy resins.

Trifluoromethanesulphonic acid can also be produced in this way.

Various derivatives of naphthoquinonediazides are still widely used for resists based on novolak resins, although these are fairly old fashioned photoinitiators. Photolysis produces a soluble indene-3-carboxylic acid by

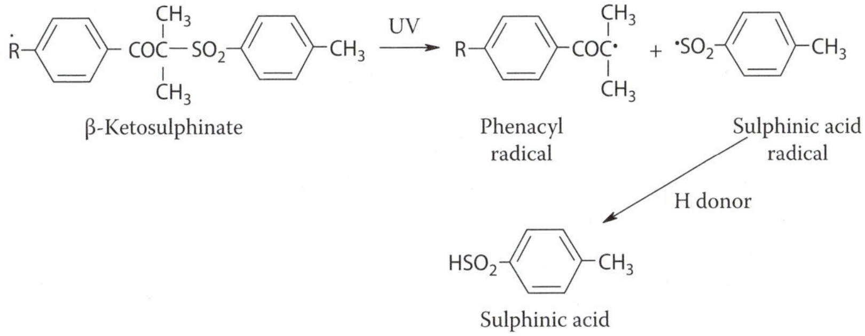

*Figure 2.24* Beta-scission and the photogeneration of a sulphinic acid.

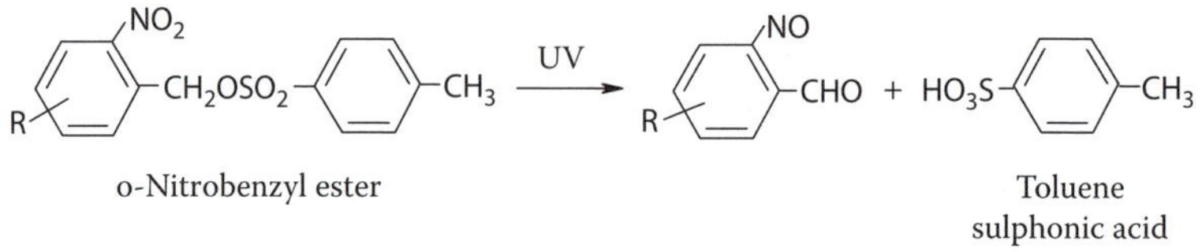

**Figure 2.25** Acid generation from an o-nitrobenzyl ester.

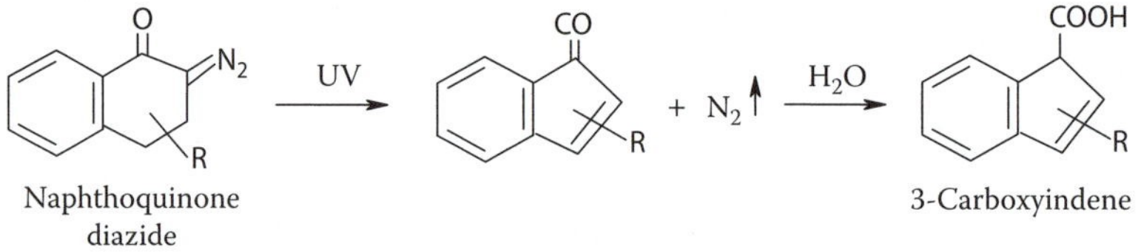

**Figure 2.26** Photodecomposition of naphthoquinone diazide.

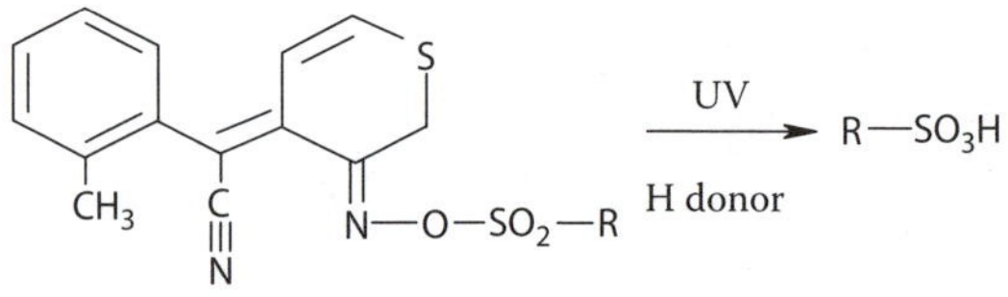

**Figure 2.27** Acid generation from an oximinosulphonate.

the elimination of nitrogen on irradiation followed by reaction with water, often from the substrate.

The 3-carboxyindene derivative that is produced in the exposed area, shown in Figure 2.26, is readily soluble in dilute alkali wash to produce an image via initial exposure through a mask.

More recent work has centered on the scission of oximinosulphonates to provide sulphonic acids. These materials show high thermal stability and high efficiency and provide non-ionic, halogen free acids, as outlined in Figure 2.27.

The Irgacure PAG series of photoacid generators (32–34),[18] based on the above scheme, will produce a variety of sulphonic acids and respond to a wide UV spectrum.

Photoinitiators that efficiently generate radicals and may produce acids as a by-product include the trichloroacetophenones (Section 2.2) and the trichloromethyl-S-triazines (Section 4.1.6), both of which can produce hydrogen chloride. In these cases the chlorine radical that is produced from the original scission process may abstract a hydrogen atom from an oligomeric donor molecule to produce hydrogen chloride.

Cationic photoinitiators produce very strong Brønsted acids that are capable of opening epoxy rings. Cationic chemistry is examined in Chapter 7.

## 2.7.8 Photo-base generation

Photo-base generation is a small specialty area that is developing some commercial interest.[17]

Type I scission of an aminoketone such as a derivative of Irgacure 907 can lead to the production of a tertiary amine as shown in Figure 2.28.

Photoinitiators such as the alkylaminoacetophenones (10–15), which have amino substitution on the α-carbon, will produce an alkylamino radical on irradiation. In the presence of a strong hydrogen donor such as a thiol, there is the potential to produce a tertiary amine. The amine produced can then be used to open epoxy groups in a ring opening polymerization or catalyze the thiol/isocyanate reaction for 2K polythiourethane formulations in auto clear coatings. The free radicals that are produced are not involved in this type of polymerization. Recent developments by Ciba have introduced the photolatent base PLA-1 based on this type of structure.[19] The substituent groups R on the nitrogen will modify the strength of the base that is produced. Substituents on the aryl structure will adjust the absorption wavelength, usually to provide UVA absorption.

Stronger bases can be provided by the photogeneration of amidines based on substances such as diazabicyclononene (DBN). A similar protecting aryl group is used that undergoes scission and, in this case, uses an intermolecular hydrogen transfer to deliver the base.

A benzylic structure in Ciba PLA-2, Figure 2.29, provides the steric hindrance of a protecting group and the photoinitiator is a relatively weak base. Photoscission produces two radicals, and hydrogen transfer from the alpha carbon leads to the formation of a double bond on the nitrogen and the production of the strong amidine base.

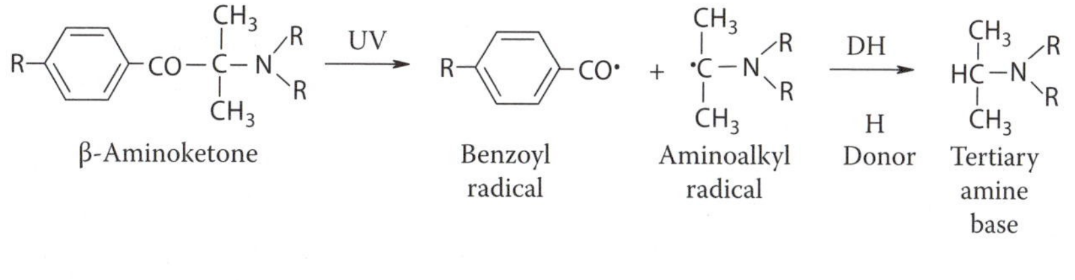

*Figure 2.28* Alpha-scission and photo-base generation with a hydrogen donor.

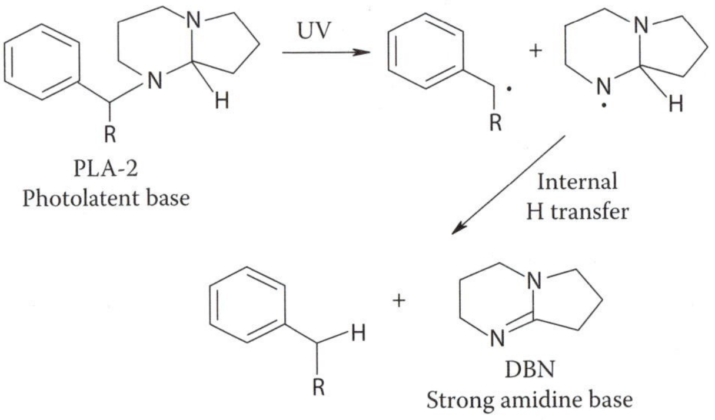

*Figure 2.29* Photogeneration of an amidine base from PLA-2.

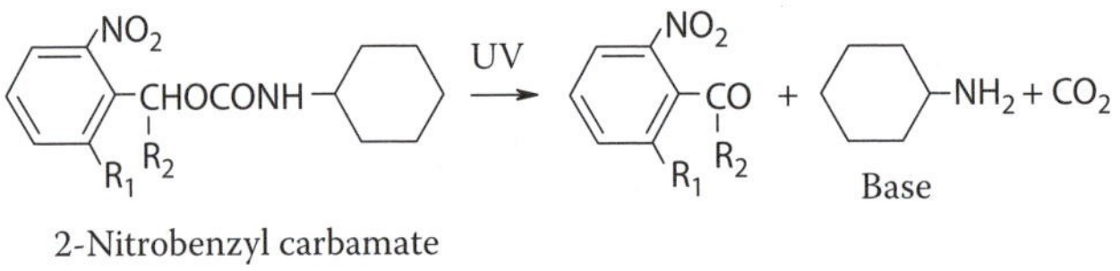

*Figure 2.30* Production of a primary base from 2-nitrobenzyl carbamates.

Carbamates protected by the 2-nitrobenzyl group or similar groups will also produce a base by scission under UV irradiation (Figure 2.30).

In this case, a primary base is generated, which in most cases is less effective as a catalyst than a tertiary base. The substituents $R_1$ and $R_2$ are required to provide a good quantum yield and stability for the reaction.

## 2.7.9   Anthracene peroxy radicals

Under UV, anthracene derivatives can form unstable endoperoxides that decompose to form radical species. In a recent development, thioxanthone and anthracene have been combined to form a photoinitiator, TX-A,[20] that forms a similar peroxide under UV but only in the presence of oxygen, as shown in Figure 2.31. TX-A has an absorption spectrum with two peaks that reflect both thioxanthone (max. 380–400 nm) and anthracene (max. 420–440 nm). During exposure, the longer wavelength of the anthracene is photobleached. The rate of polymerization of TX-A without an amine is higher than that of the standard combination of thioxanthone/amine. TX-A shows very efficient cure in air and is relatively inactive under nitrogen.

The apparent necessity of oxygen for the production of radicals with TX-A suggests an unorthodox mechanism, differing from either Type I

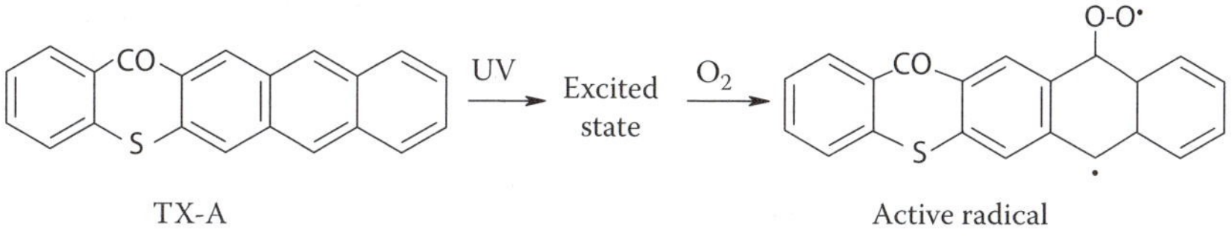

TX-A                                   Active radical

*Figure 2.31* Hybrid radical generation from thioxanthone-anthracene. (From D.K. Balta et al., 2007. *Macromolecules* 40, 4138–4141. With permission.)

or Type II. This probably involves the formation of a peroxide radical on the anthracene and its associated aryl radical. This is supported by the fact that TX-A is an efficient photoinitiator for styrene formulations and must therefore have a very short triplet lifetime, unlike Type II thioxanthones.

This is an academic study that has not yet been developed commercially, but it illustrates the continuing efforts of both academia and industry to find new, improved solutions in the technology of UV curing.

## 2.8   The thiol-ene photopolymerization

The thiol-ene stepwise photopolymerization of thiols and a variety of unsaturated compounds produces polysulphides, a different class of polymer that can bring alternative properties to an application.[21–25]

The polymerization involves the stepwise addition of a thiol to an alkene, preferentially initiated by a radical source that may be UV controlled. A sulphur-centered thiyl radical is first produced. Thiols are excellent hydrogen donors that can interact with radicals from a Type I photoinitiator and with excited triplet states of Type II photoinitiators.

From a Type I photoinitiator:

$$PI + UV \rightarrow Rad^{\cdot} \qquad RSH + Rad^{\cdot} \rightarrow RS^{\cdot} + Rad\text{-}H$$

From a Type II photoinitiator:

$$RSH + PI + UV \rightarrow RS^{\cdot} + PI\text{-}H \text{ (ketyl radical)}$$

Propagation:

$$RS^{\cdot} + CH_2 = CHR_1 \rightarrow RSCH_2C^{\cdot} HR_1$$

$$RSCH_2C^{\cdot} HR_1 + RSH \rightarrow RSCH_2CH_2R_1 + RS^{\cdot}$$

The thiyl radical adds to the unsaturation to produce a carbon-centered radical. This alkyl radical abstracts a hydrogen atom from a second thiol molecule to produce another thiyl radical that continues the polymerization process.

Using multifunctional thiols such as trimethylolpropane tris(3-mercaptopropionate) and difunctional or multifunctional "enes," polymeric networks with a wide variety of mechanical and physical properties can be created. Both flexible elastomers and extremely hard polymers can be produced by varying the stoichiometry. Any type of monomer with an "ene" function, including unsaturated esters and ethers, conjugated dienes, etc., will react. Acrylates and methacrylates are best utilized with lower levels of thiol. In particular, bis-norbornene oligomers are much more reactive than vinyl or allyl ethers, which again are more reactive than alkenes and acrylates. Norbornene responds well to relatively low intensity UV and low light dosage, using only one tenth the energy requirements of a corresponding acrylate polymerization.

The thiol-ene reaction is not subject to oxygen inhibition. Any peroxy radicals that are formed readily abstract a hydrogen atom from the thiol to regenerate a thiyl radical, effectively undergoing chain transfer and continuing the polymerization process. Thiols, like tertiary amines, act as oxygen scavengers.

The thiol-ene process also provides low shrinkage of the order of 3–5%, since it starts from an inherently low viscosity medium, and gelation occurs only close to full cure. Excellent adhesion can be achieved, which is probably related to the low shrinkage. The monomer conversion is very high, and very rapid cure speeds can be achieved. Excellent cross-link density can be achieved along with good depth cure. Thick polymers up to one centimeter are relatively easy to produce. Although the starting materials are odorous, the final polymers that are produced can be very low-odor materials.

Both Type I and Type II photoinitiators can be used as a UV radical or triplet source. Tertiary amines are not required. The addition of 2% of a simple photoinitiator such as Darocur 1173 (1) or DEAP (16) will be suitable for most applications. It is possible to achieve cure without using a photoinitiator as a radical source. Under high-intensity short wave UV, thiols will cleave to produce sulfur-centered radicals, and very thick, optically clear plastics have been produced in this manner.

One serious disadvantage of the thiol-ene process is its susceptibility to dark reaction. Formulations of this nature are a little thermally sensitive and chemically reactive, so gelling may occur and the shelf life of such formulations is limited.

Although the UV-initiated thiol-ene reaction has yet to find more general industrial use, it has been used for conformal coatings in the electronic sector, adhesives, sealants, and some printing applications.

## *References*

1. Vesley, G. F. (3M) Mechanism of the photodecomposition of initiators. (DEAP, HAPs, α-halo acetophenones, benzoin ethers, oximes, ketals, AAAPs.) *J. Rad. Curing.* 1986, Vol. 13, No. 1, 4–10.
2. Schnabel, W. (Hahn-Meitner Inst. Berlin) Mechanistic and kinetic aspects concerning the initiation of free radical polymerization by thioxanthones and acyl phosphine oxides. *J. Rad. Curing.* 1986, Vol. 13, No. 1, 26–34.
3. Decker, C. (ENSC Mulhouse). UV Curing Chemistry: Past, Present and Future. RadTech Europe 1987, Munich, Conference.
4. Rist, G. A. Borer, K. Dietliker, V. Desobry, (Ciba) J. P. Fouassier, and D. Ruhlmann. (ENSC). Sensitisation of amnoketone photoinitiators. (beta-scission). *Macromolecules.* 1992, 25, 4182–4193.
5. Dietliker, K. and J. V. Crivello. (1998) in *Photoinitiators for free radical, cationic and anionic photopolymerization*, 2nd edition, Vol. 3 in the series Chemistry and Technology of UV and EB Formulation for Coatings, Inks and Paints (Bradley, G., Ed.), John Wiley and Sons/SITA Technology, London, chapter IV. Figure 26, page 13.
6. Reich, W., C. Glotfelter, K. Saaa, H. Bankowsky, E. Beck, M. Lockai, and R. Noe. (BASF Corp.) 2000. Acylphosphine oxides as photoinitiatiors not only for pigmented coatings. *RadTech NA. Conf. Proc.* 545–559.
7. Berner, G. R. Kirchmayr, J. Puglisi, and G. Rist. *J. Radiat. Curing,* 1979 (6), 2.
8. Hulme, B. E. and J. J. Marron. *Paint Resin,* 1984 (54), 31.
9. Fink, M., W. Schnabel, S. Schneider, F. Seitz, and Q. Q. Zhu. *J. Photochem. Photobiol. A: Chem.* 1991 (59), 255.
10. Encinas, M. V., E. A. Lissi, (Univ. Santiago) A. Zanocco, L. C. Stewart, and J. C. Scaiano. Photochemistry of alkyl esters of benzoylformic acid. *Can. J. Chem.* Vol. 62, 1984, 386–391.
11. Jonsson, S., J. Hultgren, P. E. Sundell, M. Shimose, J. Owens, K. Vaughan, and C. E. Hoyle. (Fusion UV, Univ. Southern Mississippi.) Photocopolymerization initiated by N-substituted maleimides and electron donor olefin combinations. A photoinitiator free UV curable system. *RadTech Asia, Conf. Proc.* 1995. 283–295.
12. Clark, S., C. E. Hoyle, S. Jonsson, F. Morrel, and C. Decker. (Univ. Southern Mississippi.) Photoinitiator efficiency of N-functional aliphaticmaleimides. *RadTech NA. Conf. Proc.* 1998. 177–181.
13. Millar, C. W., S. Jonsson, C. E. Hoyle, C. Hasselgran, T. Haraldsson, L. Shao. (Univ. Southern Mississippi.) *RadTech NA. Conf. Proc.* 1998. 182–188.
14. Cavitt, T. C., E. Hoyle, C. Nguyen, K. Viswanathan, and S. Jonsson. (Univ. Southern Mississippi) *RadTech NA. Conf. Proc.* 2000. 785–794.
15. Davidson, R. S. 2001. Maleimides: Do they have a role in radiation curing? *PRA RADNews* 36, 19–23.

16. Dietlker, K., U. Kolczak, G. Rist, and J. Wirz. 1996. *J. Amer. Chem. Soc.* 118, 6477.
17. Frechet, J. M. J. 1992. Photogeneration of acid and base. *Pure and Appl. Chem.* 64(9), 1239–1248.
18. Photoacid generators for microlithography. 2006. Ciba Specialty Chemicals. Data sheet.
19. Dogan, N., H. Klinkenberg, L. Reinerie, D. Ruigrok, P. Wijmands. (Akzo) K. Dietliker, K. Misteli, T. Jung, K. Studer, P. Contich, J. Benkhoff, and E. Stizmann. (Ciba). 2006. Fast UV-A Clearcoat. *RadTech Report*, March/April, 43–52.
20. Balta, D. K., N. Arsu, Y. Yagci, S. Jockusch, and N. J. Turro. 2007. Thioxanthone-anthracene. *Macromolecules*. 40, 4138–4141.
21. Jacobine, A. F., D. M. Glaser, and S. T. Nakos. (Loctite Corp.) Norbornene functionalised resins as substrates for thiol-ene polymerizations. *RadTech Eu. Conf. Proc.* 1989. 465–473.
22. Woods, J. G. and A. F. Jacobine. (Loctite Corp.). 1992. Non-acrylate photopolymer systems: UV photocuring studies of norbornene/thiol resins. *RadTech NA. Conf. Proc.* 173–182.
23. Rakas, M. A. and A. F. Jacobine. (Loctite Corp.) Non-acrylate photopolymer systems: Physical properties of UV cured films of norbornene-thiol resins. *RadTech NA, Conf. Proc.* 1992. 462–473.
24. Hoyle, C. E., T. Clark, T. Y. Lee, T. Roper, B. Pan, H. Wai, H. Zhou, and J. Lichtenhan. (Univ. Southern Mississippi.) Thiol-Enes: Fast curing systems with exceptional properties. *RadTech Eu. Conf. Proc.* 2005. 289–293.
25. C. E. Hoyle. (Univ. Southern Mississippi.) Nanostructured thiol-ene photopolymerized networks. Proc. PRA Cost and Performance Conference. Manchester (UK). 2006. Paper 9.

# chapter three

# Academics unlimited

The structure of a photoinitiator is designed to produce radicals under the influence of UV light. For commercial use, this process has to be as efficient as possible and most photoinitiators presently in use provide excellent cure. There are, however, many deactivating processes that inhibit the various photophysical and photochemical steps along the road to radical production. Academic studies have developed a good understanding of these processes and offer a means for the intimate study of the efficiency of a photoinitiator, remote from the normal testing procedures in widespread use industrially.

For those interested in a better understanding of the development of photoinitiators, a simple examination of these steps in the radical process chain can give an insight into the effects of various molecular substitution patterns on the efficiency of a photoinitiator and lead to some understanding of how and why the various molecules have been commercialized.

The excited singlet state that is first produced by absorption of UV energy can relax back to the ground state by fluorescence decay. In most cases, the singlet lifetime is extremely short (0.1–2 ns) and there is usually a fast intersystem crossing (isc) to form the excited triplet state[1] (see Figures 3.1 and 1.8).

Radical production is most often from the triplet state, which leads directly to radicals via the Type I scission process. The triplet state of Type II photoinitiators, in the presence of a tertiary amine, rapidly produces a charge transfer complex (CTC, exciplex) via electron transfer from the amine. This is followed by fast proton transfer to produce radicals, of which the radical arising from the tertiary amine is the most reactive. In the absence of an amine, alternative hydrogen atom donation from alcohols, ethers, thiols, etc., can lead directly to radical production, but this is generally a less efficient process.

The triplet state can also lose energy by phosphorescence decay, but, more importantly, can be deactivated by oxygen and by monomers, losing its ability to produce radicals. In particular, strong triplet quenchers such as styrene or unsaturated polyesters can render a photoinitiator almost useless. This quenching effect will depend on the lifetime of the triplet state, allowing time for interaction, and on the rate of quenching by the various species. In practice, the concentration of a monomer species is

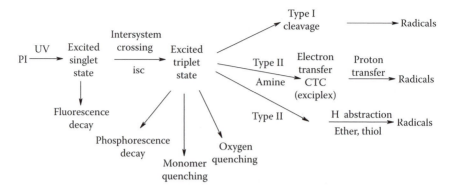

*Figure 3.1* Reactive and deactivating processes in the production of radicals.

much higher than that of oxygen in the system, and for most purposes, oxygen quenching of the triplet state can be ignored. In general, Type I photoinitiators have short triplet lifetimes that allow little time for quenching reactions. Type II photoinitiators have longer triplet lifetimes and the triplet state is easily quenched by monomers. The Type II photoinitiators are therefore much less efficient in the presence of styrene, etc.

There are numerous ways of studying the reactions of photoinitiators, each bringing a different outlook to the processes of initiation, and many academics have their own styles of working. Time-resolved laser spectroscopy on a nanosecond time scale, using laser pulses, can follow the generation of singlet and triplet states and measure the lifetimes and rate constants of the various processes by recording subsequent absorption/wavelength patterns. Similarly, the reaction of amine synergists with photoinitiators can be studied, showing the transfer of an electron and the production of a ketyl radical from the resulting absorption spectra of the intermediates.

## 3.1  Triplet lifetimes and monomer quenching reactions

A table of triplet lifetimes, monomer quenching rates, and the rates of scission (Table 3.1) gives an interesting perspective on Type I photoinitiators.[2,3]

The benzil ketal, Irgacure 651 (16), and benzoin ether (21) have extremely short triplet lifetimes (<0.1 ns) and very fast rates of cleavage to produce radicals. This makes them very efficient in the presence of strong triplet quenchers such as styrene, where little time for quenching reactions before cleavage occurs. The phosphine oxides such as Lucirin TPO (22) are also effective in these formulations. The hydroxyacetophenones, Darocur 1173 (1) and Irgacure 184 (3), have longer triplet lifetimes (30 ns) and lower

*Table 3.1* Photophysical Data of Type I Photoinitiators[2,3]

| Photoinitiator | Triplet lifetime $\tau^{-9}$ sec | MMA quenching rate $kq^{-6}$ (l/mol/sec) | Rate of cleavage $k\alpha^9$ sec$^{-1}$ |
|---|---|---|---|
| Darocur 1173 | 30 | 250 | 0.7 |
| Irgacure 184 | — | 300 | 1 |
| Irgacure 651 | <0.1 | — | 10 |
| Benzoin ether | <0.1 | — | 10 |
| TPO | <1 | — | >1 |

*Source:*  Data from J. P. Fouassier and J. F. Rabek, *Radiation Curing in Polymer Science & Technology, Vol. II. Photoinitiating Systems,* Elsevier, Essex, UK, 1993; J. V. Crivello and K. Dietliker, *Photoinitiators for Free Radical, Cationic and Anionic Photopolymerisation,* Chichester, UK: Wiley/Sita, 1998.

rates of cleavage. These factors make the 2,2-dimethyl-2-hydroxyacetophenones (HAPs) less efficient than Irgacure 651 or the phosphine oxides in styrene-based formulations.[4]

Comparative triplet quenching rates of MMA (methyl methacrylate) and styrene indicate the relative difficulty of curing formulations containing strong triplet quenchers:

MMA      800 l/mol/sec x 10$^{-6}$
Styrene    4800 l/mol/sec x 10$^{-6}$

## 3.2   Modification of hydroxyacetophenones

In the development of the HAPs, attempts to make them more efficient by substitution on the aryl group led to little success. Electron donors in the p-position will red shift the absorption to longer wavelengths and, in theory, allow more efficient use of the stronger mid range outputs of the UV light from a mercury lamp.

Substituted hydroxyacetophenone ($R_1$ = H for Darocur 1173)

In Table 3.2, methoxy, methylthio and dimethylamino substitution at $R_1$ showed that the photophysical data of the HAP had been changed significantly, affecting the triplet lifetimes, the rate of cleavage and the quantum yield, as well as the expected wavelength of absorption.[5]

Cure speed was measured in an acrylate formulation[6] and correlates well with the rate of cleavage and the yield.

*Table 3.2* Substituted Hydroxyacetophenones

| $R_1$ | $R_2$ | Absn. nm | Triplet lifetime $\tau^{-9}$ sec | Rate of cleavage $K^{-9}$ sec$^{-1}$ | Quantum yield | Cure speed m/min. |
|---|---|---|---|---|---|---|
| H | OH | 244 | 30 | 0.7 | 0.3 | 41 |
| $CH_3O$ | OH | ca 270 | 12 | 0.087 | 0.38 | 27 |
| $CH_3S$ | OH | 305 | 4000 | 0.002 | 0.004 | <5 |
| $(CH_3)_2N$ | OH | ca 320 | 3300 | 0.003 | 0.011 | <5 |
| H | $OCOCH_3$ | 250 | 2300 | — | — | — |

*Source:* Adapted from W. Schnabel et al. *J. Photochem,* 1980, 92, 225; J. Ohngemach et al. *RadTech Eu. Conf. Proc.* 1989, 639–657.

Methoxy substitution at $R_1$ [comparable to Irgacure 2959 (4) which has p-hydroxyethoxy substitution] improves the photo DSC reactivity a little, with a lower triplet lifetime even though the rate of cleavage is diminished. The absorption, however, shows little improvement with respect to picking up stronger outputs from a mercury lamp. Both methylthio and dimethylamino substitution shifts the wavelength toward 300–320 nm, allowing the stronger 313 nm UV output to be used. However, the reactivity in both cases drops very significantly, following the strongly reduced rates of cleavage. This is no doubt due to the much longer triplet lifetimes, allowing monomer quenching, etc. These factors show that a good absorption at particularly strong UV output lines is not the only factor that influences the reactivity of a photoinitiator.

Similarly, esterifying the hydroxy group at $R_2$ to acetoxy has a deactivating effect on the carbonyl group, giving a much longer triplet lifetime that leads to monomer quenching and poor cure.

## 3.3   *Alkylaminoacetophenones and wavelength selection*

Alkylaminoacetophenones were developed at a time when the UV curing of highly pigmented films was a difficult problem, and attempts to produce novel photoinitiators were aimed at moving the absorption to longer wavelengths and improving cure speeds. A cyan pigment has a strong UV absorption up to 320 nm and can best be cured by photoinitiators absorbing above this wavelength. Titanium dioxide white pigment absorbs strongly up to 400 nm, and it is essential to provide absorption at 400–420 nm to achieve good cure for white inks, with the 404 nm output of the mercury lamp being especially important.

Full substitution of the alpha carbon atom (2,2-dimethyl) on a hydroxyacetophenone is essential to provide an efficient initiating species.

*Table 3.3* Substituted Alkylaminoacetophenones

| $R_1$ | Absorption nm | Triplet lifetime $\tau^{-9}$ sec | Rate of cleavage $K^{-9}$ sec$^{-1}$ | Quantum yield |
|---|---|---|---|---|
| H | 242 | 1 | 1 | 1 |
| $CH_3O$ | 269 | 0.4 | 2.5 | 1 |
| $CH_3S$ | 303 | 10 | 0.1 | 0.88 |
| $(CH_3)_2N$ | 323 | 2000 | 0.005 | 0.014 |

*Source:* Adapted from J. P. Fouassier et al. *Eur. Polym. J.* 28(6); 593–594; K. Dietliker et al. *RadTech Eu. Conf. Proc.* 1987.

2-hydroxyacetophenone, without the 2,2-dimethyl substitution, is ineffective as a photoinitiator. A similar substitution pattern applies to the alkylaminoacetophenones; the main difference being that the more strongly electron donating nitrogen group on the alpha carbon brings a much faster reactivity, from more efficient scission, than the hydroxyl group.[7]

Substituted alkylaminoacetophenone ($R_1$ = $CH_3S$ – for Irgacure 907)

The unsubstituted molecule (H at $R_1$) shows good reactivity but the absorption is still in the short wave UV. Table 3.3 shows that methoxy substitution also provides excellent reactivity but gives little change in absorption. Methylthio substitution (Irgacure 907) shows good reactivity with a short triplet lifetime, and the improvement in extending the wavelength to 303 nm is significant. Dimethylamino substitution at $R_1$ gives a much longer triplet lifetime and a poor rate of cleavage, which leads to very low reactivity, even though the wavelength shift is positive. Confirmation of the reactivity in a cyan ink showed that methylthio substitution (Irgacure 907) gave a cure speed of 130 m/min against 20 m/min for the unsubstituted (H) initiator.

In a similar comparison, Irgacure 907 (2,2-dimethyl-2-morpholino-4'-(methylthio)acetophenone) cured at 130 m/min against < 10 m/min for 2,2-dimethyl-2-hydroxy-4'-(methylthio)acetophenone, illustrating the increased reactivity that nitrogen (morpholino) substitution on the alpha carbon brings relative to hydroxyl.[8]

Different types of nitrogen substitution on the alpha carbon (morpholino for Irgacure 907) have little effect on the wavelength of maximum absorption. Assessment of how the different types of nitrogen group on the alpha carbon would influence the reactivity of the photoinitiator has been made. The morpholino group was chosen in this instance since it showed nearly twice the cure rate in a cyan ink compared with other nitrogen-containing groups such as di-(methoxyethyl)amine, N-methyl piperazine and piperidine.

Further developments along similar lines led to the commercialization of Irgacure 369 (12).

Irgacure 369

Different substitutions on the aryl group and on the alpha carbon were examined to find the most efficient combination for use in pigmented systems. The methylthio substitution on the aryl group, which led to odor problems with Irgacure 907, was changed to morpholino. This gave another red shift in absorption to 320 nm, making the initiator much more efficient for curing cyan inks[9] as it could now pick up the 313 nm UV output more efficiently and did not have the low reactivity associated with a dimethylamino group on the aryl. The nitrogen on the alpha carbon substitution was changed from morpholino to the more electron-donating dimethylamino group, which gave another leap in cure speed.

In addition, the typical 2,2-dimethyl substitution on the alpha carbon was examined. Larger alkyl groups had little effect on the wavelength of absorption maximum but seemed to broaden the absorption and extend the "tail" toward the visible, allowing improved cure in pigmented systems. Irgacure 907 had been shown to be capable of curing white inks to some degree, but changing the 2,2-dimethyl substitution to 2-butyl-2-ethyl gave a tenfold improvement in cure due to this "tailing" effect (Figure 3.5). It became clear that bulkier groups on the alpha carbon instead of the standard 2,2-dimethyl extended the tail of the absorption significantly toward the visible.[9]

For Irgacure 369, the 2-ethyl-2-benzyl substitution was twice as effective as 2-ethyl-2-vinyl in extending the absorption above 400 nm, even though the absorption at this wavelength is very small. Compared with Irgacure 907, Irgacure 369 gives much higher cure speeds in a white lacquer, illustrating the effect of the absorption showing a broader pattern and tailing into the visible range above 400 nm (see Figures 2.14 and 3.2). Unfortunately, there is also a detrimental yellowing effect with Irgacure 369 after cure.

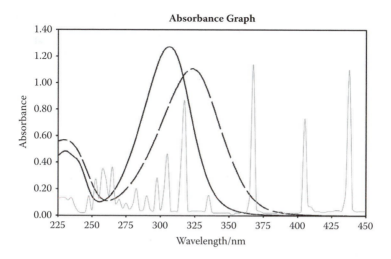

***Figure 3.2*** Wavelength tailing of Irgacure 907 (to 375 nm) and 369 (to 410 nm).

The alkylaminoacetophenones, Irgacures 907, 369, and 379, are easily sensitized by small amounts of thioxanthones, which enables very high cure speeds to be obtained. The energy transfer process is discussed in Section 3.5 and also in Chapter 5.

## 3.4   Phosphine oxides: Reactivity and solvolytic stability

Phosphine oxides were developed from esters of acylphosphonic acids, which show some photoinitiating properties, although relatively poor. The phosphine oxides have a low absorption in the ultraviolet-visible spectroscopy (UV-Vis) around 370–420 nm. Their photobleaching properties have made them prime photoinitiators for titanium dioxide white coatings and inks. They have very short triplet lifetimes and are also used for styrene based formulations and glass fiber-reinforced polyesters. Alpha cleavage gives a substituted benzoyl radical and a phosphinyl radical, the latter being very reactive but also very sensitive to oxidation.

2,4,6–Trimethylbenzoyldiphenyl phosphine oxide (TPO)

The phosphine oxides can be regarded as esters where the CO–PO bond is relatively easily hydrolyzed by any nucleophilic species such as water, alcohols, or amines. Ortho substitution on the aryl carbonyl group is essential to provide solvolytic stability. While o-chloro and o-methoxy groups also provide stability and give a similar reactivity, a p-methyl group was shown to give improved quantum yield of scission. Synthetic demands led obviously to the mesitoyl or 2,4,6-trimethylbenzoyl structure of the phosphine oxides being the most efficient.

Shielding of the carbonyl group by ortho-substituents improves the solvolytic stability and substitution of both ortho-positions gives maximum stability in a formulation.[10] This is relatively easily demonstrated by measuring the half life of the material in a mixture of methanol and water:

| | |
|---|---|
| benzoyl | <1 hour |
| 2-methylbenzoyl | ca 40 hours |
| 2,4,6-trimethylbenzoyl | >1 year |

Phosphine oxides give excellent depth cure but relatively poor surface cure, since the phosphinyl radicals are very sensitive to oxygen. Surface cure can be improved by the addition of tertiary amines, which scavenge oxygen, but the solvolytic stability effect has to be taken into account, allowing for nuclephilic hydrolysis by any basic materials.

The addition of methyldiethanolamine (MDEA) to a formulation containing phosphine oxides will lead to only a few days' stability before the photoinitiator is destroyed. The "olamines" are particularly reactive in this situation. Unsubstituted tertiary amines and less basic amines such as the aminobenzoate esters may be less reactive toward hydrolysis and can be used to improve surface cure, but care is needed.

## 3.5   Benzophenone and thioxanthone triplet reactions

Benzophenone and thioxanthone undergo similar reactions under UV irradiation, producing first an excited singlet state that may decompose by fluorescence decay. The singlet lifetime is very short and usually converts rapidly to a triplet excited state that leads on to radical formation (see Figure 3.3). Scission does not occur with these aryl ketones and either an electron transfer or hydrogen atom transfer can occur, depending on the donor material.

The rates of production, quenching, and transfer of the various species[11] are

| | |
|---|---|
| Oxygen quenching of triplet, | $3 \times 10^6 \, M^{-1} \, s^{-1}$ |
| Monomer quenching of triplet, $k_q$ | $6 \times 10^7 \, M^{-1} \, s^{-1}$ |

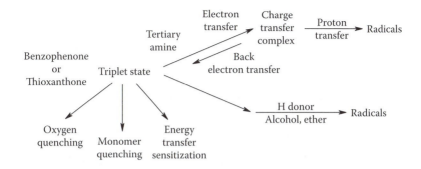

*Figure 3.3* Triplet reactions of benzophenone and thioxanthone.

| | |
|---|---|
| Energy transfer sensitization, | $6 \times 10^7$ M$^{-1}$ s$^{-1}$ |
| Electron transfer from amine, $k_e$ | $3 \times 10^9$ M$^{-1}$ s$^{-1}$ |
| Hydrogen atom donation from alcohol | $4 \times 10^6$ M$^{-1}$ s$^{-1}$ |

Aryl ketones have long-lived triplet lifetimes (150–800 ns). Strong triplet quenchers such as styrene show a high quenching rate compared with methyl methacrylate; aryl ketones such as benzophenone and thioxanthone derivatives cannot be used in styrene-based formulations. Compared with the high rate of electron transfer from an amine, monomer quenching is relatively slow, and monomers such as MMA and acrylates (other than styrene) have little effect on the triplet state in the presence of an amine despite their high concentration.

$$K_e \text{ [Amine]} \gg k_q \text{ [M]}$$

Similarly, oxygen quenching of the triplet excited state can be disregarded, since the oxygen concentration is generally low compared with that of the monomer during the light absorption process.[12] Oxygen inhibition from radical scavenging is a separate issue (see Section 4.7).

In the presence of a tertiary amine, a very fast electron transfer from the lone pair on the nitrogen to the excited triplet state occurs, producing a charge transfer complex (CTC or exciplex) in high yield (see Figure 3.4). This transfer can be deactivated by back electron transfer, but the subsequent proton transfer from the amine within the CTC is also very rapid and the former deactivating effect is very small. This process from the excited triplet through the CTC in the presence of an amine rapidly produces radicals with high efficiency. The reactivity of the tertiary amine in producing radicals has been shown to be inversely proportional to the ionization potential or basicity of the amine.

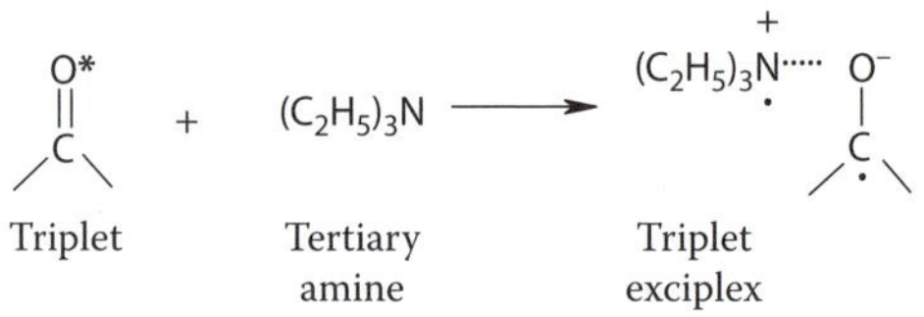

*Figure 3.4* Tertiary amine and excited triplet interaction.

Tertiary amines with the lowest ionization potential give the highest conversion rates.[13] Similar studies with a variety of olamines show that dimethylpropan-2-olamine is 50% more reactive than methyldiethanolamine (MDEA), which, in turn, is more reactive than dimethylethanolamine.[14] Hydrogen atom donation as well as electron abstraction may be responsible for the higher reactivity of the olamines. The aminobenzoate esters such as ethyl p-dimethylaminobenzoate are even more effective than the simple tertiary alkylamines or alkanolamines.

Alternatively, in the presence of hydrogen donors such as alcohols, ethers, thiols, etc., no charge transfer occurs to form a CTC, and direct hydrogen atom transfer from the donor to the carbonyl group occurs. This is a less efficient reaction than that of an amine. The result, in both cases, is that a ketyl radical, which is relatively unreactive, is produced from the photoinitiator, and the reactive radical comes from the donor, whether it is the amine or the alcohol, ether, etc.

Flash photolysis in various formats using pulsed light and measuring the responding absorption patterns can give some insight into the generation of triplet states and their subsequent reactions with tertiary amines and hydrogen donors.

Flash photolysis of benzophenone photoinitiators in nitrogen saturated isopropanol leads to the generation of a weak absorption pattern at 540–560 nm. This corresponds to a transient species, the ketyl radical, generated by the hydrogen atom transfer from the isopropanol. In hexane solution, the same transient pattern is much weaker, since hexane is a very poor hydrogen donor. In the presence of oxygen, no transient is formed in isopropanol since the triplet state is efficiently quenched by the oxygen. Both hydrogen donation and oxygen quenching have similar rates of reaction. With the addition of increasing concentrations of triethylamine to benzophenones in nitrogen-saturated isopropanol, an enhanced transient absorption is formed with a small red shift. This suggests the formation of the radical anion (triplet exciplex) through electron transfer from the amine. The addition of tetracyanoethylene, a radical trap, drastically reduces the transient formation and indicates confirmation of a charge transfer state process via electron transfer. Nitrous oxide can be used as an electron trap with similar results.

In the presence of suitable photoinitiators, benzophenone and thioxanthones can act as sensitizers, passing on energy to provide high speed photoinitiation.[15,16] The rate of transfer of energy is very high and can make sensitization a very efficient process (see Section 5.4). Energy transfer depends on the energy of the triplets involved. The triplet energy of the sensitizer, BP or TX, has to be a little higher than that of the acceptor initiator. Sensitization of alkylaminoacetophenones such as Irgacure 907 and Irgacure 369 is commonly associated with a thioxanthone such as isopropylthioxanthone. The relative triplet energies are

Speedcure ITX $\quad E_T = 61.4$ kcal/mol
Irgacure 907 $\quad E_T = 61$ kcal/mol
Irgacure 369 $\quad E_T = 60$ kcal/mol

Long wave UV energy at 380 nm is readily absorbed by the thioxanthone and swiftly transferred to the AAAP (Irgacure 907), which then produces radicals via scission. The small amount of thioxanthone (0.2–0.5%) does not require an amine; it simply acts as a catalyst, passing on energy and returning to the ground state. The larger amount of Irgacure 907 (3%) rapidly develops a high radical count without necessarily absorbing energy itself (screened by pigments). An increase in cure speed of over three times that of Irgacure 907 alone can be achieved,[17] and even greater advantages are shown with Irgacure 369.

2% Irgacure 907 alone $\quad$ 20 m/min
2% 907 plus 0.01% ITX $\quad$ 30 m/min
2% 907 plus 0.25% ITX $\quad$ 50 m/min
2% 907 plus 0.5% ITX $\quad$ 70 m/min

Benzophenone also has a high triplet energy ($E_T = 69.1$ kcal/mol) and is an excellent sensitizer in its own right. Unfortunately, BP absorbs UV in the short wave around 250 nm and cannot be used as a sensitizer in pigmented media, but similar sensitizing systems using benzophenone could be used in clear coatings.

Where the triplet energy of the sensitizer is less than that of the photoinitiator, energy transfer cannot occur. Darocur 1173 ($E_T = 67$ kcal/mol) and Irgacure 651 ($E_T = 66.2$ kcal/mol) have triplet energies higher than that of Speedcure ITX and no sensitization effect occurs when these two are combined with ITX.

## 3.6  Substituted benzophenones

Substituted benzophenone

The photophysical and photopolymerization data of several substituted benzophenones[18,19] are detailed in Table 3.4. The photoinitiators and their structures listed below refer to numbers 1 through 6 in Table 3.4.

| | | |
|---|---|---|
| 1. benzophenone, (BP) | $R_1 = R_2 = H$ | |
| 2. 2,4,6-trimethyl BP | $R_1 = 2,4,6$-trimethyl | $R_2 = H$ |
| 3. 4-methoxy BP | $R_1 = CH_3O$ | $R_2 = H$ |
| 4. 4-phenyl BP, (PBZ) | $R_1 = $ phenyl | $R_2 = H$ |
| 5. 4-phenylthio BP, (similar to BMS) | $R_1 = PhS$ | $R_2 = H$ |
| 6. 4,4'-diphenoxy BP, (DPB) | $R_1 = PhO$ | $R_2 = PhO$ |

The most electron-donating substitution in the above list of products, No. 5 (4-phenylthio BP, similar to Speedcure BMS) produces the greatest red shift in absorption to 311 nm, followed by No. 4 (4-phenyl BP, PBZ) at 284 nm, and No. 6 (diphenoxy BP) at 280 nm. The molar absorbance, $E_{max}$, of No. 5, with the longest wavelength, changes little from that of benzophenone. A large increase in molar absorbance comes from the di-substituted

*Table 3.4* Photophysical Data for Substituted Benzophenones

| $P_1$ | Absn. nm | Absorbance $E_{max}$ l.M$^{-1}$ cm$^{-1}$ | Triplet lifetime $\tau^{-9}$ sec | Rate of e trans. $K_e^{-9}$ M$^{-1}$s$^{-1}$ | MMA quenching $K_q^{-6}$ l.M$^{-1}$s$^{-1}$ | Polym. rate m/min |
|---|---|---|---|---|---|---|
| 1. BP | 249 | 18,700 | 160 | 1.3 | 66 | 22 |
| 2. tri Me | 244 | 14,400 | 790 | 0.6 | 8 | 25 |
| 3. MeO | 277 | 15,100 | 130 | 2 | 150 | 19 |
| 4. PBZ | 284 | 22,200 | 630 | 0.2 | 0.05 | 34 |
| 5. (BMS) | 311 | 18,200 | 230 | 1.2 | 2 | 33 |
| 6. DPB | 280 | 29,500 | 150 | 1.5 | 180 | 30 |

*Source:* Adapted from J. P. Fouassier et al. *Eur. Polym. J.* Vol. 27, No. 9, 991–995, 1991; J. P. Fouassier et al. *Polymer Commun. (Polymer Reports).* 1990, Vol. 31, 418–421.

4,4'-diphenoxybenzophenone, No. 6, with an increase of over 50% of that of benzophenone, but the red shift in wavelength is not as pronounced. The 2,4,6-trimethyl substituted benzophenone, No. 2, has a reduced molar absorbance but the reactivity is slightly higher than benzophenone. The least effective of these materials is 4-methoxybenzophenone, No. 3, which has a lower molar absorbance and absorbs at 277 nm.

There is little correlation among triplet lifetimes, monomer quenching rates and polymerization rates. Triplet lifetimes of all these Type II photoinitiators are very long (130–790 ns) compared with the Type I benzil dimethyl ketal (Irgacure 651) at < 0.1 ns. Monomer quenching rates in MMA and acrylates do not affect the reactivity, since the electron transfer rate from an amine is three orders higher and a charge transfer complex is rapidly formed. In styrene, the monomer quenching rates are much higher and compete with electron transfer more successfully.

The reactivity in a typical acrylate is mostly dependant on the ability to pick up energy either from the rather weak output at 254 nm or the stronger UV output at 313 nm of a mercury lamp. The latter output matches the absorption of Speedcure BMS, No. 5, almost perfectly, while 4-methoxybenzophenone, No. 3, absorbs at 277 nm, which lies midway between these output lines and makes the photoinitiator unable to absorb either output efficiently.

Polymerization rates in an acrylate system indicate much higher reactivity from the three most highly substituted benzophenones, No. 4 Speedcure PBZ, No. 5 Speedcure BMS, and No. 6 diphenoxy benzophenone. In practice, the reactivity increases in the following order, although this depends on the testing regime and is a bit subjective.

BP < Speedcure PBZ < Speedcure BMS < diphenoxybenzophenone, DPB

DPB has not been commercialized due to its very poor solubility profile, <<0.5% in most monomers, although it is one of the most reactive of the substituted benzophenones.[20] Of the other substituted benzophenones, PBZ is generally regarded as an alternative when a little extra boost in cure speed is required, although its relatively poor solubility profile needs to be kept in mind.

4,4'-Bis(dimethylamino)benzophenone (Michlers ketone, MK) is a very activated benzophenone and shows very high reactivity. Unfortunately, Michlers ketone also shows some carcinogenic effects and is no longer used. The bis(diethylamino) counterpart, EMK, although less reactive than MK, still shows excellent reactivity but suffers from comparison on safety grounds and also produces strong yellowing on cure.

## 3.7   Substituted thioxanthones

Thioxanthones absorb in the long wave UV at 360–420 nm and are used in pigmented formulations where the absorption of a pigment in the short- and mid-range UV often screens that of the photoinitiator. The thioxanthones can pick up the 366 nm and, more importantly, the 404 nm output of a mercury lamp, providing more efficient cure for UV inks. Thioxanthones, like the benzophenones, have long triplet lifetimes of the order of 190 ns, which can lead to quenching reactions, but the rate of electron transfer from an amine $k_e$ is extremely fast at $10^9$ $M^{-1}$ $s^{-1}$. This competes effectively with most monomer quenching rates $k_q$ of $10^6$ $M^{-1}$ $s^{-1}$. The formation of a charge transfer complex via electron transfer from an amine, which is very fast, effectively governs the efficiency of radical production.

The structure of the amine used with the thioxanthone (and Type II photoinitiators in general) has a significant effect on the reactivity of the system.[21] Electron transfer and radical production varies inversely with the ionization potential, or basicity, of the amine (Table 3.5).

In n-butyl methacrylate using 0.1% 4-propoxythioxanthone as a photoinitiator, the secondary amines with higher ionization potentials give the lesser performance. Of the three tertiary amines, the conversion doubles as the ionization potential decreases from 8.1 to 7.4.

Commercial production of thioxanthones is most efficient with substitution in the 2-position.

2-CTX                      2-ITX                  2,4 - DETX

Commercial thioxanthones

*Table 3.5* Reactivity of the Amine

| Amine 0.10% | Ionization potential | Conversion % |
|---|---|---|
| Piperidine | 9.76 | 0.192 |
| Dicyclohexylamine | 9.2 | 0.004 |
| Diethylamine | 8.4 | 0.175 |
| Diethylmethylamine | 8.1 | 0.304 |
| Triethylamine | 7.85 | 0.583 |
| Tri-n-butylamine | 7.4 | 0.658 |

*Source:* Data from N. S. Allen et al. *Eur. Polym. J.* 24(5). 435–440, 1988.

2-Chlorothioxanthone (CTX), 2-isopropylthioxanthone (ITX)[22] and 2,4-diethylthioxanthone (DETX) all absorb at 383–386 nm and have similar reactivity. Solubility is a hindrance with CTX, and ITX has become the industry standard. DETX has a little better solubility but otherwise there is little to choose from among these three products.

An acetyl group (CH$_3$CO-) in the 2-position has a deactivating effect and gives a much less reactive photoinitiator with an absorption at 376 nm.[23]

R = Alkyl, alkoxy
acetyl, aryl
halo, hydroxy
carboxy, etc.

Substituted thioxanthones

In general, electron donor substitution in the 2- or 4-position, and electron acceptors in the 1- and 3-positions, will all produce a red shift in absorption to longer wavelengths. Reactivity usually follows the pattern of the red shift, increasing with wavelength. Deactivating electron acceptors will have the opposite effects to the above pattern, producing a blue shift to shorter wavelengths and a reduced reactivity. Many types of substitution patterns on the thioxanthone have been made, generally improving the reactivity by electron donors in the 2-position (Table 3.6).

For methoxy substitution,[24] the reactivity follows the wavelength maximum with an improvement over ITX and CTX for 2- and 4-methoxythioxanthone.

2-MeOTX > 4-MeOTX > 3-MeOTX > 1-MeOTX

*Table 3.6* Substituted Thioxanthones

| TX | Absn. nm | Ext. coeff. log E | Conversion % | Cure speed m/min | MEK rubs |
|---|---|---|---|---|---|
| ITX | 383 | 3.84 | — | 5.8 | 80 |
| CTX | 385 | 3.68 | — | 6 | 90 |
| 1-MeO | 370 | 3.81 | 0.33 | — | — |
| 2-MeO | 395 | 3.77 | 0.53 | — | 110 |
| 3-MeO | 363 | 3.82 | 0.35 | — | — |
| 4-MeO | 387 | 3.88 | 0.48 | 11.5 | 130 |
| 3,4-diMe-2-PrO | 385 | 4.03 | 1.1 | — | — |
| 1-Cl-4-PrO | 387 | 3.91 | — | 13 | 230 |

3-Methoxythioxanthone gives a blue shift in absorption to 363 nm with respect to ITX, and shows a corresponding low reactivity. 1-methoxythioxanthone also shows a deactivating effect, and this is in part due to intramolecular hydrogen bonding of the methoxy (and also a 1-methyl) group with the carbonyl group, deactivating the triplet state. The addition of methyl groups to the activating 2-alkoxy group can also have a beneficial effect and 3,4-dimethyl-2-propoxythioxanthone[25] shows a higher extinction coefficient and increased conversion despite the slightly reduced red shift at 385 nm.

A chlorine atom in the 1-position, as opposed to a methyl group, can have an activating effect on the reactivity of a thioxanthone.[26] The photoinitiator Speedcure CPTX was developed along these lines.

1-Chloro-4-propoxythioxanthone (CPTX)

1-Chloro and 3-chlorothioxanthones undergo some degree of photo dechlorination under UV, producing chlorine and 1-thioxanthyl radicals.[27] This form of radical production occurs in the absence of a tertiary amine, although the rate of polymerization due to the production of chlorine radicals is only about half that of ITX/amine. The conjugation of the chlorine and the carbonyl, together with the inductive effect, weakens the carbon–halogen bond and scission may occur, although other mechanisms have also been proposed. A by-product of this reaction, in the presence of amines, is the formation of hydrogen chloride. Some of the photodecomposition pathways are outlined in Figure 3.5.

1-Chloro-4-propoxythioxanthone (CPTX) shows a photo dechlorination of around 40% on its own with a relative insensitivity toward oxygen. In the presence of a tertiary amine, CPTX shows more significant advantages in cure speed, hardness, and cross-link density. Some of this higher reactivity comes from the better light absorption properties of CPTX, which shows a secondary peak at 312 nm, able to respond to the 313 nm output of the UV lamp, and a red shift in long wave absorption to 387 nm.

Thioxanthone carboxylic esters have been assessed as sensitizers for the cyclo-dimerization of dimethylmaleimide photopolymers (see Figure 3.6, Table 3.7).[28] Energy transfer from the excited thioxanthone was measured as the amount of energy required to reach a No. 7 step wedge for the photopolymer.

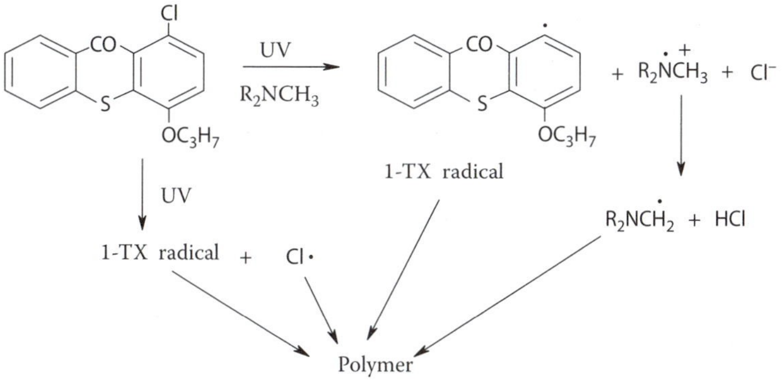

***Figure 3.5*** Photodecomposition of CPTX.

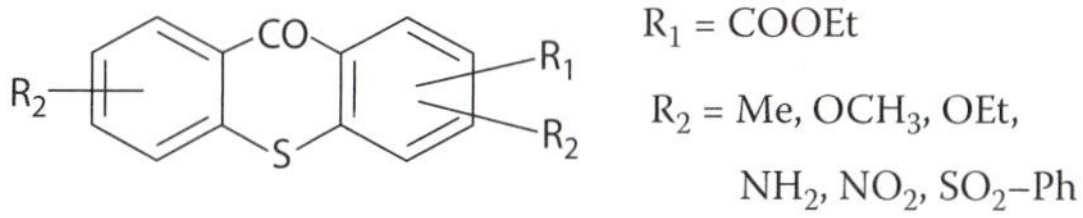

$R_1$ = COOEt

$R_2$ = Me, OCH$_3$, OEt,

NH$_2$, NO$_2$, SO$_2$–Ph

***Figure 3.6*** Substituted thioxanthone esters.

***Table 3.7*** Photochemical Data and Sensitivity Response
of Thioxanthone Carboxylate Esters

| TX sensitizer | | Wavelength | Absorption | ET | Sensitivity |
|---|---|---|---|---|---|
| $R_1$ | $R_2$ | λ max. nm | l.M$^{-1}$ cm$^{-1}$ | kcal.M$^{-1}$ | mJ.cm$^{-2}$ |
| None | None | 376 | 6200 | 63.1 | 22 |
| 1-COOEt | none | 383 | 6450 | 63.1 | 11 |
| 1-COOEt | 3–OEt | 368 | 6100 | 64.3 | 22 |
| 1-COOEt | 3–SO$_2$Ph | 398 | 5550 | 59.5 | 11 |
| 1-COOEt | 3–NH$_2$ | 333 | 11000 | 60.7 | 55 |
| 1-COOEt | 3–NO$_2$ | 407 | 4300 | 54.8 | 55 |
| 2-COOEt | nil | 377 | 5750 | 63.1 | 22 |
| 3-COOEt | nil | 397 | 6050 | 58.3 | 22 |
| 3-COOEt | 7–Me | 403 | 6050 | 58.4 | 22 |
| 3-COOEt | 7–OCH$_3$ | 416 | 6350 | 54.8 | 34 |
| 4-COOEt | nil | 386 | 7750 | 61.9 | 11 |

*Source:*  Adapted from K. Meier and H. Zweifel. *J. Photochem.* 35 (1986), 353–366.

3-Aminothioxanthone-1-carboxylate showed very high absorption, but the wavelength was blue shifted to 333 nm and the sensitizer is less responsive to the high-pressure mercury lamp that is used for photopolymers. 3-Nitrothioxanthone-1-carboxylate showed a good absorption at 407 nm but very low triplet energy. Both these sensitizers were the least effective in cyclodimerization, requiring a large dose of energy. All the other carboxylate esters showed satisfactory sensitization for polyimide cycloaddition. For efficient energy transfer in a commercial photoinitiating system, a triplet energy above 61 kcal/mol for the thioxanthone is required, and for this the 1-carboxylate ester is most efficient (63.1 kcal/mol) coupled with its longer wavelength absorption at 383 nm. The 4-carboxylate ester (386 nm, $E_T$ 61.9 kcal/mol) would also be satisfactory, but many of the other sensitizers do not have sufficiently high triplet energies to be suitable for efficient use as a sensitizer. None of these carboxylate esters have been commercialized.

## 3.8   Novel photoinitiators

Recent developments in thioxanthone compounds have uncovered a novel thioxanthone–anthracene compound[29] that takes up oxygen under UV light to produce a peroxy radical. Thioxanthone–anthracene shows good reactivity in air but is almost inactive under nitrogen. The anthracene forms a peroxy radical and associated aryl radical via a hybrid mechanism that is neither Type I nor Type II. The scheme is illustrated in Section 2.6. Thioxanthone-anthracene is, as might be expected, almost insoluble in any medium and would be difficult to incorporate into an application.

Thioxanthone-anthracene is one of many novel photoinitiators introduced by a vibrant academic sphere over many years. Much of the recent work has centered on particular applications such as photo-latent acid and base for imaging and automotive industries (Sections 2.7 and 4.1.6), excursions into silicon-based photoinitiators,[30] photopolymers and initiators for the expanding biomedical sector, novel cationic salts that do not release benzene (Section 7.3 and 7.5) and photo-latent organotin compounds.[31] The addition of electron-donating groups on the aryl to extend the wavelength and improve reactivity is well documented (see the earlier sections of this chapter) and has been studied in depth. Novel photoinitiators that do not rely on the modification of standard molecules continue to be discovered by the fertile minds of many academic schools as well as innovative industries.

An alternative method for extending the wavelength is to extend the conjugation of the carbonyl group, as shown by the dibenzylidene ketones (Section 4.5.2), which red shift the absorption from 250 nm of benzophenone up to 400 nm and have been used for water soluble photo-stencils with the addition of sulphonic acid groups. This concept has recently been taken one step further, incorporating a triple bond yne group instead of the double bond ene. 1,5-Diphenyl-1,4-diynone,[32] DPD, is an extended benzophenone that has an absorption red shifted to a broad, strong peak at 300–340 nm, is better able to pick up the 313 nm and a little of the 366 nm output of a mercury lamp.

DPD

In ethoxylated monomers where hydrogen abstraction can occur, DPD, without an amine, was shown to be twice as reactive as the standard benzophenone/amine combination. In addition, the concentration of DPD for optimum performance was only one quarter of that of benzophenone. However, further extension of the absorption by introducing electron-donating 4-aryl substituents such as –OMe, –SMe or –NMe2 led to lower reactivity.[33] DPD is a Type II photoinitiator that will perform without an amine in suitable monomers or oligomers, and the reactive species was shown to be a carbon centered monomer radical from hydrogen abstraction.

The same concept has been applied to sulphonium salts to try to extend the absorption to stronger lamp output lines. Diphenyl(phenylethynyl) sulphonium hexafluorophosphate,[34–36] PES, gives an extended absorption up to 370 nm.

PES

The reactivity of PES, compared with the standard mixed sulphonium salts, was reduced, but the epoxy group conversion was similar at 86%. The addition of dibutoxyanthracene, DBA, as a sensitizer improved reactivity to that of the standard salt. Surprisingly, the thioxanthone ITX had a similar improved sensitizing effect, whereas ITX has little influence on standard sulphonium salts. Photoproducts included diphenyl sulphide and phenyl(phenylethynyl) sulphide, but no benzene was detected. Similar phenyl(phenylethynyl)iodonium salts have been made.[37] These again showed reduced reactivity compared to the basic diphenyliodonium salt but lead to much improved response when coupled with long wave sensitizers.

Perhaps coming from the study of silicon chemistry, which can produce some photoinitiating properties, germanium compounds have recently been explored, with mono- and bis-acyl germanes showing much promise. Bis(benzoyl)diethyl germanium,[35,38–40] BBG, will cleave to give a benzoyl and a benzoylgermyl radical, both of which are reactive.

BBG

In theory, the latter radical, when part of the growing polymeric chain, could cleave further, giving a second benzoyl radical that would enhance the polymeric structure. BBG absorbs in the visible at 420 nm and has the advantage of photobleaching, which is ideal for composites. BBG in a dental application under a Bluephase LED lamp showed 30% improvement over Irgacure 819 and camphorquinone. This no doubt stems in part from its absorbance which, at the 438 nm mercury line, is much greater than that of the above photoinitiators. BBG is also stable in acidic dental primer formulations.

Developments in the biomedical field include the addition of polyether side chains to the bis-phosphine oxide Irgacure 819[41–43] to produce some hydrophilic nature and better compatibility with biomedical polymers.

In the expanding world of UV curing, new applications are constantly being studied, and no doubt new developments in photoinitiators will continue to service these requirements.

The development of novel photoinitiators is not always straightforward. Successful commercialization depends on many factors, not least of which is low toxicity, good solubility, ease of formulation, ease of handling, and cost effectiveness. Reactivity may not always be a prime requisite. The simpler thioxanthones such as 2-isopropylthioxanthone and 2-chlorothioxanthone can be made in a one-step reaction. More reactive thioxanthones such as 2-alkoxy and the 1-chloro derivatives may require a two-stage reaction process, which can significantly increase the cost of production. Unless the advantages of a new photoinitiator can withstand the increased cost, the product is not likely to succeed. Cost effectiveness will also depend on the application, whether it is for a low-cost varnish or a high-end electronics use. There have been many good, novel photoinitiators that have fallen by the wayside due to the lack of cost effectiveness. Taking a leaf from the pharmaceutical industry, it could be that molecular modeling will provide one of the future pathways, predicting wavelength, absorbance, and reactivity prior to producing the sample product.[44]

## 3.9   Radical reactions

Radicals, once formed, will react with monomers to give a monomer radical that continues the process through the growing polymer chain radical to a polymer. This will be the main reaction stream as long as there is a reasonably high level of radicals being produced; a high radical count. There are other processes that radicals undergo, some of which are deactivating,

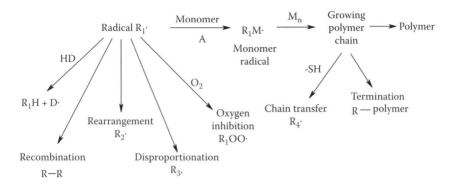

***Figure 3.7*** Radical reactions.

and others related to photo and thermal reactions that produce a variety of photoproducts (see Chapter 6, Section 6.2).

Some of these radical reactions are outlined in Figure 3.7.

### 3.9.1   Primary radical reactions

Radical reactions, whether propagating, rearranging, or terminating, will depend on the concentration of the radicals and the viscosity of the medium, allowing molecular movement. With a high photoinitiator concentration and a low viscosity formulation, the rate of the deactivating processes will increase and, for example, termination processes will produce an abundance of short chain polymers, and oxygen inhibition, such as in UV inkjet, will be problematical. Using low concentrations of photoinitiators and high-viscosity media, there will be less termination and less oxygen inhibition, and longer-chain polymers will be produced.

The prime reaction, A, with a monomer, produces the monomer radical that continues as a growing polymer chain, cross-linking, etc., to give the required polymer. Where the concentration of photoinitiator is sufficiently high, this will be the dominant process.

Reaction of any of the radicals with a hydrogen donor HD, from $R_1$ to $R_4$ including the monomer radical and the growing polymer radical, will abstract a hydrogen atom and leave a donor radical D˙. Radical transfer with donors such as alcohols, ethers, esters, etc., is generally less efficient than that of monomer addition and can be regarded as less reactive despite the fact that alternative radicals are produced.

### 3.9.2   Recombination reactions

Recombination reactions include the simple reversal of the initial scission process that produces two radicals, regenerating the photoinitiator. The two radicals, or diverse radicals, may also recombine to produce other neutral species. In the case of Irgacure 651:

The initial scission process is reversible in the sense that the benzoyl and benzyl radicals may recombine, regenerating Irgacure 651. These two radicals can also recombine in a different fashion, producing a semi-quinone structure. Two benzoyl radicals may form benzil, and the benzoyl radical may also react with the active methyl radical, produced from thermal scission of the benzyl radical, to produce acetophenone. All of these structures are neutral compounds, produced by deactivating radical reactions.

### 3.9.3   Rearrangement reactions

Rearrangement reactions can lead to the production of new, more reactive radicals in some instances, and in others can be deactivating, giving neutral products. The benzyl radical from Irgacure 651 in Figure 3.8 undergoes thermal rearrangement to produce methyl benzoate and a methyl radical, which is very reactive.

The biradical in Figure 3.9, produced from diethoxyacetophenone (DEAP) by a Norrish Type II reaction, has a fairly low reactivity and rearranges to give a neutral cyclic ether.

The benzoyl and alkyl radicals produced from the photo-scission of Darocur 1173 (Figure 3.10) can recombine to regenerate the photoinitiator or rearrange to produce benzaldehyde and acetone, both neutral species. Although both of the primary radicals are reactive in a normal polymerization, this particular minor reaction is deactivating.

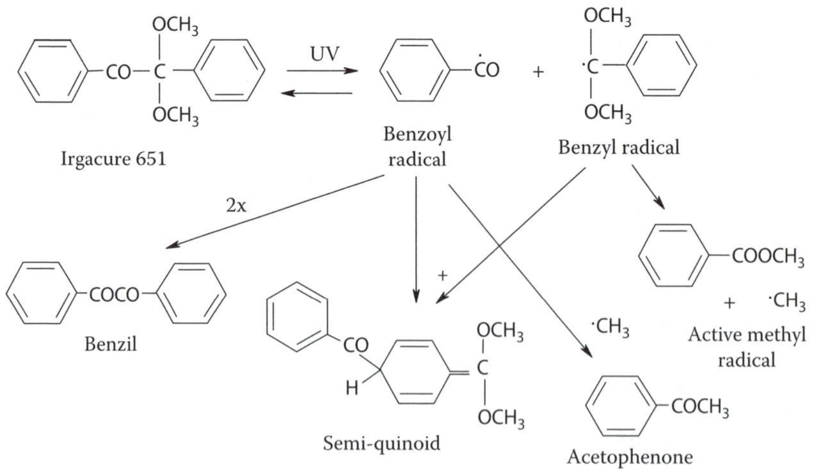

*Figure 3.8* Recombination of radicals.

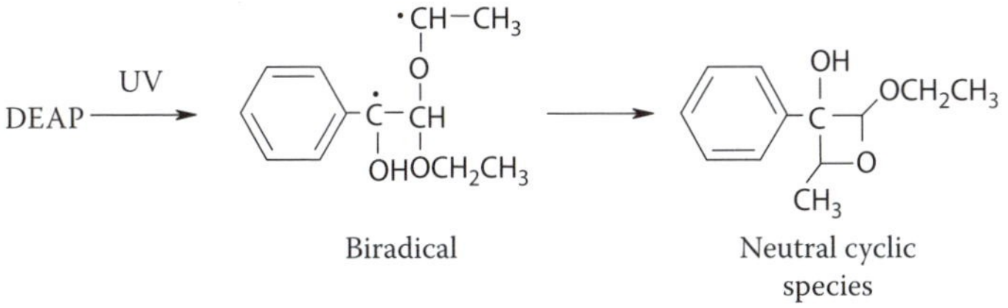

**Figure 3.9** Cyclization of a biradical.

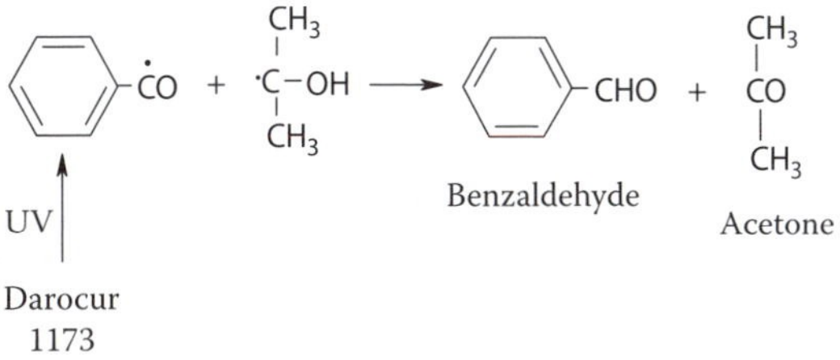

**Figure 3.10** Radical rearrangement.

## 3.9.4 Disproportionation reactions

Disproportionation can happen where two identical radicals interact to produce two neutral species; again, it is a deactivating process.

The semi-quinone radical produced from the photoreaction of 2-ethylanthraquinone, EAQ, and a hydrogen donor (Figure 3.11) can disproportionate, producing a dihydroquinone and regenerating EAQ.

The ketyl radical of Type II species such as thioxanthone can also disproportionate (Figure 3.12).

The ketyl radical produced from isopropylthioxanthone and similar Type II species usually forms a pinacol dimer (e.g., benzophenone). In the case of thioxanthone, steric considerations retard this reaction and the more likely process is that of disproportionation, where two ketyl radicals produce a hydroxythioxanthene and regenerate ITX.

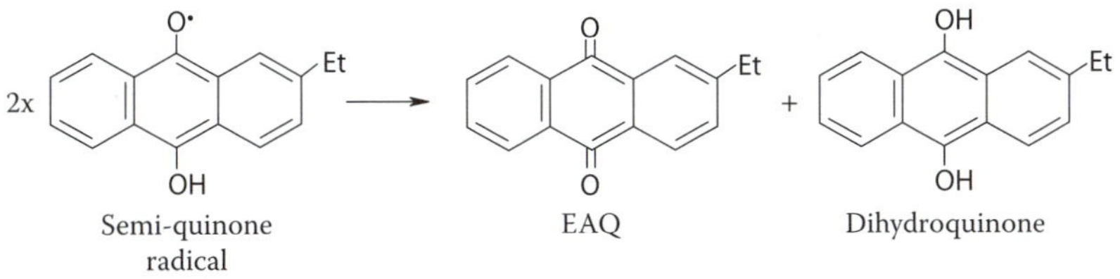

**Figure 3.11** Disproportionation reactions.

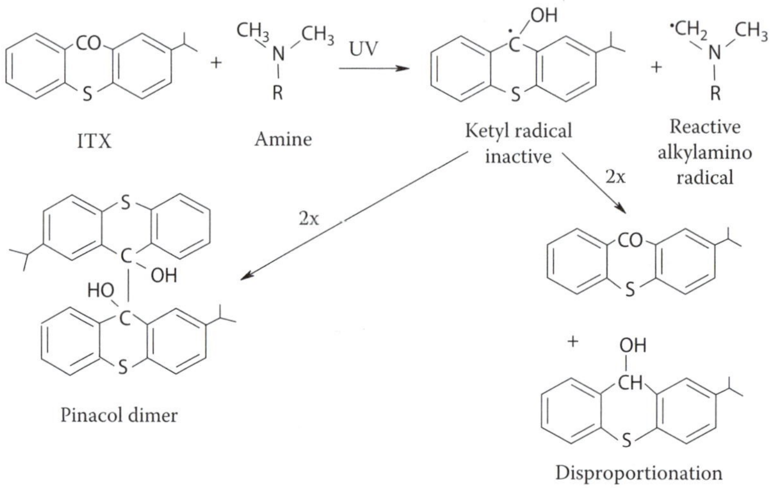

**Figure 3.12** Disproportionation of ITX.

### 3.9.5 Oxygen effects and chain transfer

Oxygen inhibition is discussed in Chapter 5, Section 5.2. Briefly, any radical species will react with oxygen within the film or more usually at the surface during cure. The peroxy radicals that are produced are inactive toward monomer reactions and do not contribute to the polymerization process. Essentially, oxygen retards the polymerization, most commonly resulting in a tacky surface.

$$R_1^{\cdot} + O_2 \rightarrow R_1OO^{\cdot} + HD \rightarrow R_1OOH + D^{\cdot}$$

The peroxide radical that is formed is inactive but can react with a strong hydrogen donor such as a tertiary amine to produce a neutral hydroperoxide. In the presence of such donors, a donor radical is formed that is sufficiently reactive to continue the monomer reaction, but otherwise, oxygen merely acts as a radical scavenger.

Chain transfer can occur in the presence of materials such as thiols, which are strong hydrogen donors. Studies have indicated that this is usually from the growing polymer chain but in theory it can occur with any radical source. A hydrogen atom from the thiol terminates the polymer growth, producing a reactive thiyl radical that starts the radical plus monomer process all over again.

$$R_1M^{\cdot} + \text{-SH} \rightarrow R_1MH + \text{-S}^{\cdot}$$

Thiols are sometimes used as chain transfer agents to limit polymer growth and give some control over the average molecular weight of a polymer.

### 3.9.6 Termination reactions

Termination reactions are simply the addition of a radical species to the radical on the growing polymer chain. Radical–radical interactions produce neutral species and end the growth of a chain. Any radical can be involved but most notably the ketyl radicals that are produced in Type II reactions, and are present in considerable concentrations, are involved in termination reactions. Termination reduces the reactivity of a formulation and decreases the molecular weight of the polymer. Examination of the polymer structure from Type II formulations will often show an end cap group that is based on the Type II ketyl radical.

### 3.9.7 Identification of radicals and excited states

Radicals can be identified chemically by reaction with other radicals. Stable radical traps form compounds that can be identified by GCMS or other means and the radical can be deduced from the structure of the compound. A well-used radical trap is based on the hindered nitroxyl radical in 2,2,6,6-tetramethylpiperidinoxyl (TMPO). UV irradiation of a photoinitiator in the presence of TMPO will produce a compound whose structure can be identified by GCMS or other means. For example, a Type I photoinitiator containing a benzoyl group, Figure 3.13, will give the following reaction:

The radical (benzoyl) can be deduced from the structure of the compound produced by interaction with TMPO. Similarly, the phosphine oxide Lucirin TPO produces a trimethylbenzoyl radical and a diphenylphosphinyl radical, both of which have been identified by trapping with TMPO.[45]

An alternative method is to use a non-polymerizable monomer such as 1,1-diphenylethylene shown in Figure 3.14. In this case, a radical adduct is formed that can abstract a hydrogen atom from a suitable donor to

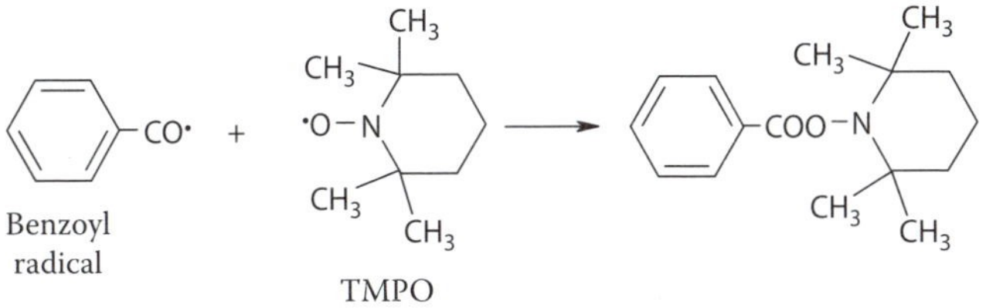

*Figure 3.13* Radical trapping with TMPO.

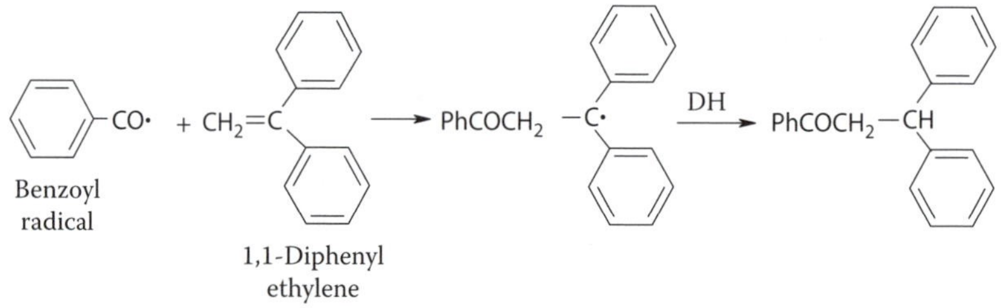

**Figure 3.14** Radical reaction with non-polymerizable monomer.

produce a stable neutral species whose structure will indicate that of the radical.

1,1-Ditolylethylene has been used in similar fashion to identify the alkylamino radical that is produced from the Type II benzophenone/amine reaction.

A similar non-polymerizable monomer is methyl 2-tert-butylacrylate (TBA), which will follow the reaction indicated in Figure 3.14. TBA has been used to identify the methylthiobenzoyl radical that is generated from the photoscission of Irgacure 907.

There are also spectrographic methods of following the excited states that lead to radical production. Electron spin resonance spectroscopy (ESR) has been used to determine excited states, identify radicals, and follow radical additions such as those above. Time-resolved ESR coupled with spin trapping with nitroso compounds can isolate radicals that are produced during photoinitiation.[46]

Nuclear magnetic resonance spectroscopy (NMR) has been used similarly, following $^1$H or $^{13}$C atoms to identify products via CIDNP (chemically induced dynamic nuclear polarization).[47] The magnetic field causes the radicals to align and produces highly resolved spectra. Singlet or triplet states can be identified and a whole variety of radical–radical interactions can be followed to deduce the numerous photoproducts that may be produced.

## References

1. Fouassier, J. P., P. Jacques, D. J. Lougnot, and T. Pilot. 1984. Lasers, photoinitiators and monomers: A fashionable formulation. *Polym. Photochem.* 5, 57–76.
2. Fouassier, J. P. and J. F. Rabek, Eds. 1993. *Radiation Curing in Polymer Science & Technology, Vol. II. Photoinitiating Systems.* Essex, UK: Elsevier. pp. 1–61.
3. Crivello, J. V. and K. Dietliker. 1998. *Photoinitiators for Free Radical, Cationic and Anionic Photopolymerisation* G. Bradley, Ed. Chichester, UK: Wiley/Sita. pp. 115–181.
4. Abadie, M. J. M. and M. Rouby. 1995. Etude comparee de la reactivite de quelques nouveaux photoamorceurs radiculaires. *Eur. Polym. J.* 31(3) 301–304. (Fr).

5. Eichler, J., C. P. Herz, I. Naito, and W. Schnabel. 1980. Laser flash photolysis investigations of primary processes in the sensitised polymerisation of vinyl monomers IV: Experiments with hydroxy alkalphenones. *J. Photochem.* 12, 225.

6. Ohngemach, J., M. Koehler, and G. Wehner. *RadTech Eu. Conf. Proc.* 1989. pp. 639.

7. Ruhlmann, D., F. Wieder, and J. P. Fouassier. 1992. Relations between structure and properties. Morpholino ketones. *Eur. Polym. J.* 28(6), 591–599.

8. Meier, K., M. Rembold, W. Rutsch and F. Sitek. 1987. In *Radiation Curing of Polymers. (Spec. Pub. No. 64)* D. Randell, Ed. Royal Society of Chemistry. p. 196.

9. Bussian, B. M., V. Desobry, K. Dietliker, H. Karfunkel, M. Kohler, and L. Misev. 1987. *Radiation Curing of Polymers (Spec. Pub. No. 64).* D. Randell, Ed. Royal Society of Chemistry. p. 163.

10. Boettcher, A., A. Henne, and M. Jacobi. 1985. *Polym. Paint. Colour. J.*, 175, 636.

11. Fouassier, J. P. and J. F. Rabek, Eds. 1993. In *Radiation Curing in Polymer Science & Technology, Vol. II. Photoinitiating Systems.* London: Elsevier. pp. 18–32.

12. Allen, N. S., E. Lam, J. L. Kotecha, W. A. Green, and A. W. Timms. 1990. Photochemistry of 4-alkylamino benzophenone initiators. *J. Photochem. Photobiol, A: Chem.* 54, 367–388.

13. Allen, N. S., F. Catalina, J. L. Mateo, R. Sastre, P. N. Green, and W. A. Green. 1989. Photochemistry of water soluble benzophenone initiators. *Polymer. Mat. Sci. Eng.* 60, 10–14.

14. Allen, N. S., E. M. Howells, E. Lam, F. Catalina, P. N. Green, W. A. Green, and W. Chen. 1988. Photopolymerisation of a water soluble benzophenone: Influence of tertiay amine. *Eur. Polym. J.* 24(6), 591–593.

15. Borer, A. V., Desobry, K. Dietliker, J. P. Fouassier, G. Rist, and D. Ruhlmann. 1992. *Macromolecules* 25, 4182–4193.

16. Fouassier, J. P., B. Graff, D. Ruhlmann, and F. Wieder. 1995. New insights in photosensitizers-photoinitiators interaction. *Progress Organ. Coat.* 25, 169–202.

17. Berner, G. R., Kirchmayer, and W. Rutsch. 1984. New photoinitiators for pigmented systems. *RadTech NA. Conf. Proc.* 5.50–5.71.

18. Fouassier, J. P. and D. Ruhlmann. 1991. Relations between structure and properties—I. Benzophenone derivatives *Eur. Polym. J.* 27(9) 991–995. (Fr).

19. Fouassier, J. P. and D. J. Lougnot. 1990. BMS, an aryl aryl sulphide derivative. *Polym. Commun. (Polymer reports)* 31, 418–421.

20. Decker, C. and K. Moussa. 1989. Diphenoxybenzophenone. *J. Polym. Sci. Polym. Lett.* 27, 347–54.

21. Allen, N. S., F. Catalina, J. Luc-Gardette, W. A. Green, P. N. Green, W. Chen, and K. O. Fatinikun. 1988. Spectroscopic properties and photopolymerisation activity of 4-n-propoxythioxanthone. *Eur. Polym. J.* 24(5), 435–440,.

22. Allen, N. S., F. Catalina, P. N. Green, and W. A. Green. Photochemistry of thioxanthones—I. *Eur. Polym. J.* 21(10), 841–848, 1985.

23. Davis, M. J., J. Doherty, A., A. Godfrey, P. N. Green, J. R., A. Young, and M. A. Parrish. 1978. The UV curing behaviour of some photoinitiators and photoactivators *J. Oil Col. Chem. Assoc.* 61, 256–263.

24. Allen, N. S., F. Catalina, B. Moghaddam, P. N. Green, and W. A. Green. 1986. Photochemistry of thioxanthones—III. *Eur. Polym. J.* 22(9), 691–697.

25. Allen, N. S., F. Catalina, P. N. Green, and W. A. Green. Photochemistry of thioxanthones—IV. *Eur. Polym. J.* 22(10), 793–799, 1986.

26. Allen, N. S., D. Mallon, W. A. Green, A. W. Timms, F. Catalina, T. Corrales, S. Navaratnam, and B. J. Parsons. 1994. Photochemistry of novel 1-halogeno-4-propoxythioxanthones *J. Chem. Soc. Faraday Trans.* 90(1), 83–92.

27. Green, W. A. and A. W. Timms. 1993. Aspects of the photodecomposition of CPTX. *Polym. Paint. Col. J.* 183(4322), 41.

28. Meier, K. and H. Zweifel. 1986. Thioxanthone ester derivatives: Efficient triplet sensitizers for photopolymer applications. *J. Photochem*, 35, 353–366.

29. Arsu, N., D. K. Balta, S. Jockush, N. J. Turro, and Y. Yagci. 2007. Thioxanthone-Anthracene. *Macromolecules* 40, 4138–4141.

30. Lalevee, J., M. Rozz, F. Morlet-Savory. B. Graff, X. Allonas, and J-P. Fouassier. *Macromolecules*. 40, 8527–8530.

31. Carroy, A. et al. (Ciba) 2009. On-demand curing of 2K-PUR with photolatent catalysts. *RadTech Eu. Conf. Proc.*

32. Seidl, B., K. Icten, and R. Liska. (Vienna University of Technology) 2005. New insights into 1.5-diphenyl-1,4-diynone as photoinitiator. *RadTech Eu. Conf. Proc.* pp. 57–65.

33. Seidl, B., K. Icten, G. Ullrich, and R. Liska. (Vienna University of Technology) 2005. New cross-conjugated photoinitiators. *RadTech Eu. Conf. Proc.* pp. 399–406.

34. Liska, R. and B. Seidl. 2005. *J. Polym. Sci. A: Polym. Chem.* 43, 101–111.

35. Liska, R., B. Ganster, C. Heller, M. Hofer, and S. Kopeinig. (Vienna University of Technology) 2007. Photoinitiators with bathochromic shift. *RadTech Eu. Conf. Proc.*

36. Hofer, M., C. Heller, and R. Liska. (Vienna University of Technology) 2007. New sulphonium salt based photoinitiators for cationic photoplymerisation. *RadTech Eu. Conf. Proc.*

37. Hofer, M. (Vienna University of Technology) 2009. Cationic photoinitiators based on phenylethynylonium salts. *RadTech Eu. Conf. Proc.*

38. Gugg, A. (Vienna University of Technology) 2009. Mechanistic investigations on germanium based photoinitiators. *RadTech Eu. Conf. Proc.*

39. Ganster, B., U. K. Fischer, N. Mosner, and R. Liska. 2008. *Macromolecules* 41(7), 2394–2400.

40. Durmaz, Y. Y., M. Kukut, N. Monszner, and R. Liska. 2009. *Macromolecules* 42(8), 2899–2902.

41. R. Liska et al. (Vienna University of Technology) 2009. Photoinitiators for biomedical applications. *RadTech Eu. Conf. Proc.*

42. Ganster, B., G. Ullrich, U. Salz, N. Mosner, and R. Liska. (Vienna University of Technology) 2007. Water soluble bisacylphosphine oxides for the photopolymerization of acidic aqueous dental formulations. *RadTech Eu. Conf. Proc.*

43. R. Liska et al. (Vienna University of Technology) 2009. Monomers and photoinitiators for biomedical applications. *RadTech Eu. Conf. Proc.*

44. Cavitt, T. B. (Abilene Christian Univ.) 2008. Modelling photoinitiators: A new walk for R & D. *RadTech NA. Conf. Proc.*

45. Baxter, J. E., R. S. Davidson, and H. J. Hageman. 1987. *Makromol. Chem., Rapid Commun.* 8, 311.

46. Baxter, J. E., R. S. Davidson, H. J. Hageman, K. A. McLauchlan, and D. G. Stevens. 1987. *J. Chem. Soc., Chem. Commun.* 73.

47. Berner, G. J., Puglisi, R. Kirchmayr, and G. Rist. 1979. *J. Radiat. Curing*, 6, 2.

# chapter four

# Commercial photoinitiators

Most of the many different types of photoinitiators are based on a range of established structures. Each of these structures brings distinct properties to its use. In both Type I and Type II photoinitiators, minor substitution patterns on these structures bring variations in physical properties, UV absorption and photochemical reactivity. The distinct groups of photoinitiators and variations in their structure are examined below.

Hydrogen donors are essential for the production of radicals with Type II photoinitiators, and the aminoalkyl radicals produced from the tertiary amines that are widely used are very efficient radicals that can also scavenge oxygen and provide increased surface cure. The reactivity of a Type II photoinitiator is therefore influenced by the structure of the hydrogen donor that is used in the formulation. The addition of tertiary amines to Type I photosystems can have a beneficial effect on the performance, often increasing surface cure, but this may be at the expense of some loss of through cure and hardness. The reactivity of a photoinitiator, among many factors, depends on the media in which it is used, and comparisons of reactivity can be made only when they are used in a standard formulation and application. Photoinitiators will often perform a little differently in other situations, so the comments that follow can be of a general nature only.

The majority of photoinitiators are available from a variety of sources under different trade names. The original source trade name is used in the text where possible. Trademarks and suppliers are listed in Table A.10.

## 4.1 Type I photoinitiators (see Tables A.1 and A.2)

### 4.1.1 Hydroxyacetophenones

R = H, alkyl, hydroxyalkoxy

Hydroxyacetophenones

The hydroxyacetophenones (HAPs) have a strong absorption in the short wave UV around 230–270 nm and, in moderate concentrations, can be used for surface cure in pigmented systems. They also have a weak absorption at longer wavelengths up to 360 nm that can access the 313 nm output of a medium-pressure lamp to provide some depth cure. This makes them very versatile and efficient photoinitiators.

The HAPs are mainly used in clear coatings, overprint varnishes, topcoats for wood, metal, plastics, and adhesives, and are also used in silicone-based coatings. They produce hard coatings and high gloss with good reactivity. The alkyl radical produced gives rise to colorless photoproducts, and their lack of long wave UV absorption and aromatic photoproducts makes them ideal for very low yellowing clear coatings and varnishes. The HAPs are less suitable for styrene formulations than the benzil ketals such as Irgacure 651 (14) due to their longer triplet lifetimes. The HAPs can also be used in water-based formulations. The high triplet energy of the HAPs makes them less suitable for fast-curing inks where pigment absorption has to be considered, since they cannot be sensitized in the long wave UV by thioxanthones. An early evaluation of photoinitiators included Darocur 1173, Irgacure 651, benzoin ether, Irgacure 184, BP, and Quantacure ITX.[1]

2,2-Dimethyl-2-hydroxyacetophenone, Darocur 1173 (1), and 1-hydroxy-cyclohexyl phenyl ketone, Irgacure 184 (3),[2] have similar reactivity and are widely used in high quality clear coatings and varnishes. For low-viscosity coatings, they need to be used in fairly high concentrations to provide surface cure. Darocur 1173, a liquid, has some volatility but the p-tert butyl version, Omnirad 102 (2), is less volatile. Irgacure 184 is a solid with good solubility in UV monomers.

The polymeric version of 1173, Esacure KIP 150 (6),[3–5] has very low volatility and will provide very low odor and low migration clear coatings, though the pure material is a gum that is difficult to handle. Solutions of KIP 150 in other liquid photoinitiators and in monomers are available (Section 4.3), and these are more easily used. An aqueous emulsion of KIP, Esacure KIP EM (106), can be used in water-based UV formulations.

2-Hydroxy-4'-(2-hydroxyethoxy)-2-methyl-propiophenone, Irgacure 2959 (4),[6] has a good reactivity similar to, if not more reactive than, Darocur 1173. Irgacure 2959 can provide very low odor and low volatility since it does not produce benzaldehyde as a photoproduct. The p-hydroxyethoxy group on the structure gives Irgacure 2959 a hydrophilic nature with a water solubility of around 2%, which makes it more compatible with and more ideally suited to water-based UV formulations such as aqueous polyurethane dispersions. This structure, however, leads to a lower solubility in standard UV monomers for oil-soluble formulations. Irgacure 2959 is a solid that possesses low odor and low volatility. It is also used in

UV-curable powder formulations and adhesives. Irgacure 2959 has FDA approval for use in food packaging and is cytocompatible with biomedical applications. The 4'-(2-hydroxypropoxy)-version, Omnirad 669 (5), performs similarly to Irgacure 2959.

Irgacure 127 (7)[7] is a bifunctional hydroxyacetophenone that displays considerably higher photoactivity than Darocur 1173. Irgacure 127 is described as a low-emission HAP, giving very low migration, and shows low volatility and low odor. Yellowing tests show Irgacure 127 to give the lowest values of all of this particular group of photoinitiators. However, Irgacure 127 has a poor solubility profile due to its higher molecular weight.

Irgacure Micro-PICS (9), which are new designs of polyfunctional HAPs built around a high molecular weight core, are presently being developed for very low emission coatings.[8]

Esacure ONE (8)[9] is a bifunctional hydroxyacetophenone similar to Irgacure 127 that shows excellent reactivity, and can provide very fast surface cure. ONE offers very low odor levels and low migration but suffers from low solubility. A liquid version, Esacure ONE 75, is a 75% solution of ONE in 25% ethozylated TMPTA and will aid formulation. Esacure ONE is FDA approved (FCN 772).

All of the hydroxyacetophenones release acetone on curing. This is mostly driven off by the heat of the UV lamps and has no real effect on application, yellowing, odor, or cure.

Blends of Irgacure 184 and benzophenone (Irgacure 500 (81)), or blends of Darocur 1173 and benzophenone (Irgacure 4665 (82)), give increased surface cure and somewhat higher cure speeds in the absence of amines than the initiators on their own, but this can be at the expense of some depth cure and hardness (Section 4.3).

## 4.1.2 Alkylaminoacetophenones

$$R_1 - \text{C}_6\text{H}_4 - CO - \underset{\underset{\text{Alkyl}}{|}}{\overset{\overset{\text{Alkyl}}{|}}{C}} - N\underset{R_2}{\overset{R_2}{<}}$$

Alkylaminoacetophenones

The alkylaminoacetophenones (AAAPs) are versatile, fast-curing photoinitiators that are much more reactive than the HAPs and have strong absorptions in the mid UV range around 280–350 nm. The high reactivity arises from the nitrogen substitution's providing higher electron density on the $\alpha$-carbon atom, making the scission process more efficient. The AAAPs are very reactive photoinitiators that are used in high-speed offset and flexo inks, UV ink-jet, etch resists, printing plates, and solder masks.

They can also be used with care for very reactive clear coatings, but they tend to lead to some yellowing, particularly in thick films.

2-Methyl-4'-(methylthio)-2-morpholino-propiophenone, Irgacure 907 (10),[10,11] absorbs at 303 nm and is a very efficient, fast curing photoinitiator. One of its photoproducts, methylthiobenzaldehyde, is a volatile liquid that gives a strong sulphury odor, immediately recognizable on curing. Irgacure 907 can be used in screen inks and white lacquers with only minor yellowing, but its main application is in resist inks. Irgacure 907FF is a granular powder version of 907 that is free flowing, devoid of lumps, generates little static, and is easier to incorporate into a formulation.

Quadracure MMMP-3 (11) is a polymeric version of Irgacure 907 using an oligomeric ε-caprolactone-thio instead of the 4'-methyl-thio substitution on the aryl group.[12] This removes the odor problem and MMMP-3 provides very low odor and low extractables, although photoproducts are produced through the scission process.

2-Benzyl-2-(dimethylamino)-4-morpholino-butyrophenone, Irgacure 369 (12),[13,14] absorbs at 320 nm and is the most reactive of the AAAP photoinitiators. Irgacure 369 is used in high speed inks such as flexo, offset litho, and UV ink-jet. It has low odor but suffers from only moderate solubility and also leads to more yellowing than Irgacure 907, particularly in thick films under long exposure times. A polymeric version of 369 is available as Omnipol 910 (15). A similar polymeric version is marketed as Quadracure BDMD-3 (13) which, like MMMP-3, employs an oligomeric ε-caprolactone substitution on the aryl nitrogen.[12]

2-(4-Methylbenzyl)-2-(dimethylamino)-4-morpholino-butyrophenone, Irgacure 379 (14),[7] a much more soluble version of Irgacure 369, is described as a low-emission amino ketone (LE-AK). Irgacure 379 has almost identical photochemical properties and reactivity to 369 and gives improved depth cure in inks. Irgacures 369 and 379 are perhaps the most reactive of all commercial photoinitiators.

Irgacures 907, 369 and 379 are easily sensitized in the long wave UV by small amounts of thioxanthone. This sensitization effect increases their reactivity significantly and leads to very high cure speeds that can be utilized in high-speed offset litho and flexo inks (Chapter 5, Section 5.4).

### 4.1.3   Benzil ketals and dialkoxyacetophenones

Benzil ketals

Benzildimethyl ketal (BK), 2,2-dimethoxy-2-phenylacetophenone,[1] (BDK, DMPA, Irgacure 651 (16)) shows very fast photochemical cleavage and fragments, along with its photoproducts, to produce a methyl radical that makes it a very efficient photoinitiator. It can, however, produce some odor and yellowing from the rearrangement of the substituted benzyl radical that is first formed from the primary scission process. The main by-product, methyl benzoate, has a strong odor and BDK cannot be used for food packaging or for applications where taint may be a problem. BDK, like DEAP, absorbs in the short-wave UV around 230–270 nm. BDK also has a significant smaller absorption up to 370 nm that allows increased response to the 313 nm output of the mercury lamp, making it fast and efficient at providing both surface and depth cure.

The very short triplet lifetime of BDK means that there is little time for adverse monomer quenching reactions to occur. BDK is very efficient in unsaturated polyester resins and styrene formulations for wood coatings and particleboard fillers. BDK has a high thermal stability that leads to excellent formulation stability. BDK, one of the most widely used photoinitiators,[15] is used in numerous applications for both inks and coatings. Combinations of BDK and other photoinitiators such as Irgaure 369 can be used to boost the cure of the darker colors and the blacks in UV inks. BDK can be regarded as a good general-purpose photoinitiator, giving high reactivity and producing good hardness and gloss.

Diethoxyacetophenone, DEAP (18), has similar characteristics to BDK, fragmenting to produce ethyl radicals; but competing intramolecular hydrogen abstraction reactions (Norrish Type II, producing a less reactive biradical) make DEAP a less efficient photoinitiator. DEAP[16] was used in wood coatings and varnishes but is being steadily replaced by the more efficient HAPs.

## 4.1.4  *Benzoin ethers*

Benzoin ethers

The benzoin ethers (BEs) (19–21) absorb at 230–270 nm and cleave to give a benzoyl radical and a less reactive benzyl radical. These ethers contain a benzylic hydrogen atom that is chemically reactive, which leads to some thermal instability and poor shelf life in formulations containing these products. The branched longer chain ethers such as the isobutyl ether (Vicure 10) provide some structural hindrance and are more stable.[1]

The benzoin ethers can produce some yellowing, but they are efficient in unsaturated polyester/styrene formulations and are low-cost materials for wood coatings and putties for particleboard.

The benzoin ethers are, in the main, being replaced by the more efficient and now more cost effective HAPs.

## 4.1.5   Phosphine oxides

R = phenyl  (TPO)

R = mesitoyl  (819)

R = ethoxy  (TPO-L)

Phosphine oxides

The phosphine oxides (POs) have a small absorption band in the long wave UV around 350–420 nm, which makes them ideally suited to cure at depth, picking up the 366 nm and 404 nm outputs of the mercury lamp, which provide better penetration. Their non-yellowing, photobleaching properties contribute to good through cure by allowing the UV light to penetrate as cure proceeds. Although the POs are yellow materials, any remaining PO in the cured coating will photobleach on further exposure to UV or daylight to give excellent whites. The POs are used for thick film, highly pigmented formulations and are particularly useful for white lacquers containing titanium dioxide. For outdoor applications, the POs can be combined with UV absorbers and hindered amine light stabilizers (HALS). These stabilizers can also be used with most other photoinitiators. They do not interfere with the curing process and may provide some synergistic advantage. The POs are also used in printing plates and glass fiber-reinforced polyesters. The POs have very short triplet lifetimes and can be used in formulations containing monomers that are strong triplet quenchers, such as styrene.

The phosphinyl radicals that are produced are very sensitive to oxygen, which leads to poor surface cure when they are used alone, and POs are not suitable for thin film curing. This inhibition problem can be overcome in thicker films by using combinations of short wave UV surface curing photoinitiators such as Darocur 1173 (1) with Lucirin TPO (22) or with Irgacure 819 (24), which gives a good balance of surface and depth cure when used in ratios of around 3:1. There are commercial blends available for this purpose, such as Darocur 4265 (84), Irgacure 1700 (86), and others.

All the POs can be sensitized to a small degree by thioxanthones. This will give a little boost to the cure speed, but comes at the risk of some yellowing from the thioxanthone. Other sensitizers such as fluorenone (79) (absorbs at 470 nm), when used in very small amounts such as 0.05% with the PO, will enable very thick coatings of several centimeters to be cured with visible light.

Phosphine oxides are sensitive to nucleophilic attack (particularly with basic materials such as ethanolamines) which causes hydrolysis and rapidly destroys the initiator. The use of "olamines" or simple tertiary alkylamines in a PO formulation will lead to a very short shelf life. Deactivated tertiary amines, such as the aminobenzoate esters that are less basic, have less effect and may be used with care in PO formulations to improve surface cure.

Diphenyl-(2,4,6-trimethylbenzoyl) phosphine oxide, Lucirin TPO (22),[17,18] is the most widely used PO, predominantly in titanium dioxide coatings, white inks, and thick film curing. Lucirin LR 8953 is a very pure form of TPO that is suitable for electronic applications and for coating optical fibers where a very clear, consistent coating is required. The reactivity and the photochemical properties of LR 8953 are similar to those of TPO.

Ethyl (2,4,6-trimethylbenzoyl)phenylphosphinate, Lucirin TPO-L (23),[19] the alkyl ester PO, is a liquid and is a little less reactive than the solid TPO. TPO-L, being a liquid, is much more easily incorporated into a formulation and can aid the solubility of other photoinitiators, providing good miscibility. TPO-L can also be used in small amounts in water-based UV formulations.

Phenyl-bis-(2,4,6-trimethylbenzoyl) phosphine oxide, Irgacure 819 (24),[20] has a much stronger absorption than TPO in the long wave UV and is around twice as efficient as TPO. The strong absorption in the visible spectrum makes Irgacure 819 ideally suited to highly pigmented media for outdoor exposure due to the photobleaching effect of sunlight that removes the instability that would come from any remaining photoinitiator. Outdoor paints have been developed using 819 for cure in sunlight. Irgacure 819 can lead to high cure speeds and very high molecular weight polymers being formed from the secondary scission of an oligomeric radical that is first formed, and 819 tends to produce higher cross-link density in a polymer (Chapter 2, Section 2.5). Low levels of Irgacure 819 can be used to cure composites and gel coats. Irgacure 819 is available as a water-based emulsion, Irgacure 819DW (107), for aqueous UV formulations. Combinations of Irgacure 819 and Irgacure 2959 are used in UV-curable powder formulations to provide depth and surface cure.

The first generation of bis acyl phosphine oxides, bis-(2,6-dimethoxybenzoyl)-2,4,4-trimethylpentylphosphine oxide,[21] BAPO-1, is used in several commercial blends of photoinitiators but are not available as a single product. Irgacure 819 has proved to be the more resilient product.

## 4.1.6   Specialties

### 4.1.6.1   Acyloximino esters

Acyloximino esters

These photoinitiators, such as Speedcure PDO (35),[22] absorb in the short wave UV around 260 nm and are very efficient photoinitiators, producing radicals from γ scission of the N–O bond. The oximino esters have a very low triplet energy (53 kcal/mol[−1]) which leads to poor thermal stability and short pot life in a formulation. They can be used for curing unsaturated polyesters and styrene as well as acrylates and have been used for glass fiber-reinforced fabrication. Speedcure PDO is easily sensitized and can improve the efficiency of visible light curing systems using dye sensitizers such as eosin in combination with amines. PDO is now used almost exclusively in imaging applications for the electronics sector.

New photoinitiators based on acyloximino esters have recently been developed for the growing electronics sector.[23,24] Modified acyloximino esters, Irgacure OXE01 (36) and OXE02 (37), are used with sensitizers for very specific applications such as color filter resists, imaging, the production of LCDs and flat screen displays, and polyimide potting. They are unsuitable for general coatings, having been developed primarily for the above applications.

### 4.1.6.2   BCIM and HABIs

2,2-Bis-(2-chlorophenyl)-4,4′,5,5′-tetraphenyl-2′H-(1,2′)-biimidazolyl (BCIM) (38)[25] is used in the electronics sector, where its reaction with a leuco dye sensitizer will produce color formation from the dye (Chapter 2, Section 2.3). BCIM can be easily sensitized by dyes such as xanthene, acridine, or crystal violet to absorb at various wavelengths in the visible spectrum and respond to laser irradiation.

The lophyl radical that is first formed from the absorption of UV energy undergoes electron transfer with leuco dyes to produce color from disproportionation of the leuco dye radical. The system is widely used for direct write laser imaging. BCIM is just one of many HABIs that can also be used as standard photoinitiators coupled with strong electron donors, such as thiols, as the coinitiator. 2-Mercaptobenzthiazole and 2-mercaptobenzimidazole are particularly effective donors and provide high sensitivity. The relatively high molecular weight of BCIM at 630 means that it will be practically non-migratable from a UV-cured coating. However, this very large molecule also means that BCIM has a poor solubility in UV monomers and is mainly used in solvent-based applications followed by a drying step, such as in thin film printing plates. BCIM is frequently used for negative photoimaging. There are several types of HABI (38–41) with substitution patterns providing absorption at various wavelengths.

### 4.1.6.3    Photoacid generators

Photoacid generators are specialty photoinitiators that are used in microlithography, photoimaging, and photoresists. Many oligomeric materials have been developed that will produce acid functionality on the polymer under irradiation. The usual requirement is to provide solubility of an area of the coating that has been irradiated and produce an image after exposure via a dilute alkali wash. Positive images are produced where the exposed area becomes more soluble, as outlined earlier. Negative images are produced where the exposed image produces a less soluble or insoluble polymer as in standard UV curing.

In other applications, photoacid generators are available that will produce acids as simple molecules, such as p-toluene sulphonic acid, to provide catalytic chemical reactions in situ when exposed (Section 2.6 and Figures 2.28 and 2.29).

The Irgacure PAG series of photoinitiators (32–34)[26] are based on non-ionic oximinosulphonate materials that will cleave to produce a variety of sulphonic acids such as propane sulphonic acid (PAG 103 and 203), octane sulphonic acid (PAG 108), and toluene sulphonic acid (PAG 121) (Figure 2.31 and Section 2.6). The PAGs have a wide spectral response that will pick up all the main output lines of the MPM lamp including the 436 nm g-line, and can also be activated more specifically by laser. PAG 203 is designed for deep UV resists at 248 nm responding to the KrF excimer laser. These are used in producing semiconductor circuits, LCDs, etch and plating resists, holograms, chemically amplified positive resists and negative resists, and in photoimaging. They can also be used in coatings based on acid hardening systems. These types of photoinitiators are very sensitive to visible light and should be handled only under yellow light conditions.

Ciba CGI-1906 and CGI-1907 are perfluoro derivatives, similar to the PAGs, designed specifically for deep UV microlithography.

Esacure 1001M (62)[27,28] will produce toluenesulphinic acid by Type I scission under extended irradiation without a hydrogen donor. Full cure as an acid hardening agent is achieved only by a subsequent thermal step. Esacure 1001M is, however, much more efficient as a Type II photoinitiator in the

presence of a tertiary amine, where it becomes a fast-curing benzophenone
with low odor and low migration properties suitable for inks and coatings.

The naphthoquinone diazides are first-generation materials that produce alkali-soluble substituted indene-3-carboxylic acid on exposure in
imaging applications. They are still used in novolak systems for resists
and printing plates (Figure 2.28).

Photoacid generators that produce halo acids as a by-product from
a halo radical include alpha-haloacetophenones, trichloromethyl-S-
triazines, and bromomethyl phenyl sulphone.

### 4.1.6.4   *Alphahaloacetophenones*

α-Haloacetophenones

The α-chloroacetophenones, such as Trigonal P1 (25) and Sandoray 1000
(26), are efficient photoinitiators that absorb in the short wave UV at
230–270 nm. They produce chlorine radicals on irradiation by β-scission.
The chlorine radical is very reactive and will initiate polymerization, but
it can abstract a hydrogen atom from the oligomer and generate hydrogen
chloride, etc. For many applications, this acid can give rise to corrosion
problems, but there are applications such as acid-catalyzed curing where
these materials can be used successfully (Section 2.6).

In a similar fashion, phenyl tribromomethyl sulphone, BMPS (31),
produces bromine radicals, which are very reactive. These may abstract
a hydrogen atom from the oligomer and go on to produce hydrogen bro-
mide. Both trichloromethylacetophenone and BMPS are fairly old-fash-
ioned photoinitiators that have been used for thin film printing plates
and photoimaging. They have been superseded by the more efficient
trichloromethyl-S-triazines.

### 4.1.6.5   *Trichloromethyl-S-triazines*

Trichloromethyl-S-triazines

The trichloromethyl-S-triazines (27–30) produce chlorine radicals on photolysis, similar to the alpha-haloacetophenones. A downside of this, as before, is the generation of hydrogen chloride by ongoing hydrogen abstraction processes. These materials tend to be used in thin film coatings, dry film resists, printing plates, and LCDs,[23] where subsequent heat treatment can drive off the acid if necessary.

The S-triazines can be used with leuco dyes such as LCV (129) for color-forming reactions to give high-stability printing plates. They can also be used with cyanine dyes as sensitizers to respond to the argon ion laser at 488 nm for the production of high-speed photopolymer plates. In this application, the use of strong electron donors, such as 2-mercaptobenzimidazole, as a chain transfer agent, produces very reactive thiyl radicals that greatly increase the sensitivity of the formulation.

Substituting the methoxyphenyl group for a methoxystyryl or dimethoxystyryl group shifts the absorbance to longer wavelengths, from 328 nm to 470 nm. The bifunctionality of the S-triazines, from the two $CCl_3$ groups, can lead to cross-linking reactions and makes the S-triazines more efficient than the haloacetophenones.

### 4.1.6.6    Photobase generators

Photogenerated base can be used in photoimaging for tone reversal, cross-linking of epoxies, Michael reactions, and polyol/isocyanate addition reactions (see Section 2.6.8). The main application, which has been recently developed, is for OEM automobile topcoats using a thiol/isocyanate base catalyzed addition reaction. Base-catalyzed UV curing can also be used in adhesives, foams, and thermosetting resins as well as in standard epoxies. Polymeric latent bases have been developed for imaging applications.

o-Nitrobenzyl carbamates can produce secondary amines under UV irradiation, but these are less efficient than tertiary amines for most purposes. Modified alpha aminoacetophenones offer latent bases that will produce tertiary amines.

Early photolatent amines, such as PLA-1 based on Irgacure 907, required the addition of a thiol to provide efficient hydrogen atom donation to develop the tertiary amine. Irgacure 907 will deliver N-alkylmorpholine as the base, but this has only a moderate basicity (pKa 8–9). This may be sufficient for some epoxy polymerizations but is insufficiently strong for Michael additions. Irradiation of these materials, with a thiol hydrogen donor, provides fast generation of the base and the subsequent polymerization of epoxies, etc., can be accelerated by a thermal step. The weaker base of the remaining photoinitiator species produces a dark reaction that continues to post-cure the resin in shadow areas.

Ciba PLA-2[29–32] is a modified version that can be sensitized by thioxanthones and will then respond to UV A, using xenon or fluorescent lights.

Due to steric hindrance, the base in the PLA-2 photoinitiator molecule is fairly inactive until photoscission removes the blocking benzyl substitution, intramolecular hydrogen abstraction takes place, and delivers a strong amine. The base that is produced from PLA-2, 1,5-diazabicyclo-[4.3.0.]-non-5-ene, DBN, is sufficiently strong (pKa 12–13) to be effective for most base catalyzed reactions.

## 4.2   Type II photoinitiators

The performance of Type II photoinitiators (see Tables A.3 and A.4.) depends not just on the structure of the initiator but significantly on the structure of the tertiary amine hydrogen donor that is used with it. The radicals produced from amines such as dimethylaminobenzoate esters are more reactive than those from simple alkyl tertiary amines in this respect (Section 4.7).

The Type II abstraction process is also sensitive to the polarity of the medium, being more efficient in nonpolar media. A Type II system containing acidic components will lead to very poor reactivity, since the acid will simply neutralize the amine and remove the hydrogen donor.

### 4.2.1   Benzophenones

Benzophenone (BP) (50) is a very cost effective photoinitiator that is widely used in varnishes and inks. Combinations of benzophenone and amine synergist have been studied in detail[33] as the basic aryl ketone-hydrogen donor reaction. Benzophenone[1] and its alkyl-substituted homologues have a strong absorption in the short wave UV range around 230–260 nm and can be used in relatively high concentrations to give good cure in low viscosity coatings. Although BP is generally used for surface cure, it has a small absorption around 330 nm that can lead to some degree of depth cure. For higher-quality coatings, acrylated amines, or oligoamines that lead to reduced "bloom" are often used, whereas simple amines such as methyldiethanolamine are used as hydrogen donors with BP for the "cheap and cheerful" end of the market.[33]

Benzophenone has a low melting point and is able to form liquid eutectic pairs with many other photoinitiators. Benzophenone also has some degree of solvolysis, aiding formulation with less soluble photoinitiators, and has excellent pigment wetting properties when used in inks.

Benzophenone is used in perfumery and has a distinctive odor that may carry through to the finished product. The low molecular weight of BP at 182 makes it very prone to migration and a more recent trend is occurring to replace BP in some applications with its alkyl substituted homologues, although these too will be prone to migration. Benzophenone can lead to some long-term yellowing.

4-Methyl substituted benzophenone, Speedcure MBP (51), has similar reactivity to benzophenone but is less volatile and less odorous.

Esacure TZT (99), a mixture of methylbenzophenone and trimethyl-benzophenone is a little more reactive than BP. Both these products are beginning to replace BP for some applications.

The ortho-ester, methyl 2-benzoylbenzoate Genocure MBB (61), is less reactive than BP but is a very cost effective photoinitiator.

A copolymerizable version of BP, Uvecryl P36 (53), is a derivative of the ortho-ester and will reduce migration and blooming in screen inks, but it has a low reactivity.

Polymeric versions of BP, such as Omnipol BP (54), Genopol BP-1 (55), Speedcure 7005 (56) and Goldcure 2700 (57), with molecular weights approaching 1000, will give practically zero migration for packaging inks.

## 4.2.2 Substituted benzophenones

Substituted benzophenones (SBPs) absorb in the mid UV range around 280–330 nm and are more reactive than benzophenone.

R = H (benzophenone)

R = phenyl (PBZ)

R = tolylthio (BMS)

Substituted benzophenones

4-Phenylbenzophenone, PBZ, Trigonal 12 (58),[34] with its mid-wave UV absorption at 283 nm, is considerably more reactive than BP and is used in inks to add some reactivity. PBZ has poor solubility that may lead to formulation instability due to crystallization if it is used in higher concentrations and can produce some yellowing from the photoproducts. A polymeric version of PBZ, Goldcure 2300 (59), is available.

The sulphur-activated BP, 4-(4'-methylphenylthio) benzophenone, Speedcure BMS (60),[35,36] absorbing at 312 nm, is low-yellowing by comparison with PBZ, and is much more soluble in UV monomers. Speedcure BMS is the most reactive of the benzophenone derivatives. In comparison, the cure speed of Speedcure BMS, when used with the most reactive aminobenzoate hydrogen donors is somewhere between Irgacure 907 and Irgacure 369 in the Type I series. Under strong short wave UV, BMS shows traces of minor cleavage at the C-S bond to give two radicals, but the Type II hydrogen abstraction is the dominant process. This minor cleavage can lead to taint from the photoproducts but there is little sign of odor.

The sulphonylketone, Esacure 1001M (62),[27] has a similar reactivity to BMS (60) and Irgacure 907 (10) when used with the aminobenzoate Speedcure EDB. In a UV clearcoat with an acrylated amine, 1001M is about twice as fast as BP. Esacure 1001M with its higher molecular weight of 482.6 provides relatively low migration. It also provides very low odor and low yellowing but has poor solubility. 1001M can be sensitized by a thioxanthone and, with the addition of 0.5% ITX, a tripling of the cure speed may be obtained.

### 4.2.3  Thioxanthones

Thioxanthones

The thioxanthones (TXs, 63–72) are a family of photoinitiators[1,37] that are used almost universally in ink formulations. Their high triplet energy can be used to sensitize other photoinitiators, particularly the alkylaminoacetophenones, such as Irgacures 907 and 369, to provide very high cure speeds. Thioxanthones absorb in the long wave UV range at 350–410 nm, which leads to very good through cure and good adhesion. Their low absorbance at these wavelengths allows light to penetrate to considerable depth and they are able to pick up the 404 nm output of a mercury lamp, which is beyond the absorption of most pigments. All the TXs are intrinsically yellow and do not photobleach. The yellow color of any unreacted TX will affect the finished product to some degree.

Thioxanthones are used in screen inks, offset litho inks and flexo inks, often with the dimethylaminobenzoate esters as hydrogen donors. Thioxanthones can also be used in white lacquers as long as concentrations are kept sufficiently low to reduce the yellowing effect, but the "whites" market has largely been taken over by the phosphine oxides.

The large-scale commercial production of isopropylthioxanthone produces a mixture of the 2- and 4-isomer, 2-,4-ITX (63). The mixture contains about 83% 2-isomer, 17% 4-isomer and traces of other isomers. As single isomers, the 4-isomer is slightly more efficient than the 2-isomer (64) but only the single 2-isomer is available in commercial quantities via an alternative process. The 1-isopropyl isomer is much less reactive due to hydrogen bonding of the alkyl group with the carbonyl structure. 2-,4-Isopropylthioxanthone (ITX) has a good solubility in most UV monomers and is available from numerous sources.

2,4-Diethylthioxanthone, Kayacure DETX (65), was primarily made in Japan for the Far East market but is now more widely available. DETX has a marginally better solubility than ITX.

2-Chlorothioxanthone, Speedcure CTX (66), was one of the first thioxanthones to be produced, but it suffers from a very poor solubility profile and has been superseded by ITX and DETX.

2,4-Dimethylthioxanthone, Kayacure RTX (67), and 2,4-diisopropyl-thioxanthone, Kayacure DITX (68), were developed in the Far East and are very similar to DETX. There is little difference in the reactivity of all of these TXs.

1-Chloro-4-propoxythioxanthone, Speedcure CPTX (69),[38–41] has only a moderate solubility but is the most reactive of the thioxanthones and works most effectively with high amine concentrations. Electron donating propoxy substitution in the 4-position provides an additional absorption band at 314 nm, which is ideally suited to pick up the 313 nm output of the mercury lamp. This increased absorption most likely contributes to the higher photoactivity of CPTX. The molar absorption of CPTX at 404 nm is also stronger than that of ITX or DETX and makes CPTX more efficient for curing white titanium dioxide coatings, although consideration has to be taken of the yellowing effects.

Under strong UV exposure, CPTX can release a chlorine radical, which may take a hydrogen atom from the oligomer and lead to traces of hydrogen chloride being formed. In standard formulations with an amine present, traces of hydrogen chloride will form a salt with the amine, reducing the amine concentration for hydrogen donation. Speedcure CPTX is finding considerable use as a sensitizer for cationic iodonium salts.

Esters of thioxanthone carboxylic acids[42] have been used as photoinitiators and sensitizers. In particular, the electron withdrawing carboxylic substitution in the 1- or 3-positions provides esters such as ethyl (1- and 3)-thioxanthone carboxylate that are very efficient sensitizers. These materials, however, have not been commercialized.

Polymeric TXs, such as Speedcure 7010 (70), Omnipol TX (71), and Genopol TX-1 (72), will give very low migration from a cured ink and are designed for food packaging applications.

## 4.2.4   Anthraquinones

Anthraquinones

Anthraquinone sulphonic acids (AQs) were used for UV-curable printing plates in the early days of UV curing.

2-Ethylanthraquinone, EAQ (77), is almost uniquely used in the electronic sector for acid resists where amines cannot be used. In these applications the binder provides hydrogen donor sites, which leads to excellent cross-linked polymerization. Anthraquinones are relatively unaffected by oxygen. Oxygen is consumed during the photo-dissociation process, regenerating EAQ by oxidation of the dihydroxyanthracene intermediate. This makes EAQ an efficient photoinitiator in the absence of an amine[43] (Section 2.6.2). Anthraquinones absorb around 280–320 nm in the mid UV range but have been replaced by the more efficient TXs for use in more conventional Type II systems for inks. Recent research work on acrylamido-substituted AQs indicates that much improved photoactivity can be achieved, but these new molecules remain in the academic sphere.

## 4.2.5   Benzoylformate esters

Methyl benzoylformate

Benzoylformate esters are excellent non-yellowing photoinitiators that absorb in the short wave UV range around 230–260 nm. An amine hydrogen donor is not required for this Type II reaction. The Norrish Type II reaction from the $\gamma$-carbon dominates the photolysis, producing a biradical. The biradical has a low reactivity toward acrylates, but it can abstract a hydrogen atom from the resin structure very efficiently, producing an oligomeric radical (photografting) that leads to excellent cross-link density and very hard coatings. These initiators have relatively low oxygen sensitivity. They are not suitable for styrene formulations due to their long triplet lifetimes.

Methyl benzoylformate, Genocure MBF (73),[7,44] responds well to urethane acrylates without the addition of an amine and is used in thick film PVC floor lacquers and ski lacquers, where a hard coating is required. The very low yellowing factor contributes to its use in wood coatings. Incorporation of an amine into the formulation with MBF may lead to some instability and poor shelf life. MBF has only a moderate cure speed.

A bifunctional benzoylformate ester, Irgacure 754 (74), gives much higher reactivity, very low odor, and lower migration for clear coatings. Irgacure 754 is one of the lowest yellowing photoinitiators, suitable for high quality clear coatings.[45,46] Very thin water-based coatings formulated with Irgacure 754 have been shown to form an efficient base coat for plastics and metals, giving improved adhesion that comes from the oligomeric radical that is essentially photografting. Irgacure 754 can also be used in waterborne dispersions.

### 4.2.6  Camphorquinone

Camphorquinone

Camphorquinone (CQ, 75), which absorbs in the near visible range with a peak absorption at 468 nm, is used in dental applications where the 438 nm output of a "blue light" can be used. This UV-VIS light contains no harmful UV and will not affect the tissue or mucous membranes during dental repair. CQ, a Type II photoinitiator,[47,48] is generally used with dimethylaminobenzoate derivatives as the hydrogen donor species. CQ, being a 1,2-dione, photobleaches, destroying the carbonyl conjugation, and the initial strong yellow color fades and produces excellent whites. CQ is, however, a relatively slow curing photoinitiator and is not sufficiently active for UV coating or ink applications.

Other 1,2-diones such as benzil and phenanthrenequinone have been used for UV curing but have low reactivity and have not been commercialized.

## 4.3  Blends of photoinitiators

A large number of blends of photoinitiators are available commercially (see Table A.5). Some of these are proprietary mixtures that comprise a "readymade package" of photoinitiators suitable for a particular application such as for an "offset ink," a "pale flexo ink," a "white ink," etc. These blends are simply added to the binders, allowing an easy solution to formulating various products. Many of the main agencies and manufacturers of photoinitiators offer their own series of proprietary mixtures.

Other blends are made to produce a liquid eutectic from two or more solid photoinitiators that will ease the problems of dissolution. Yet others provide a synergistic mixture that brings improvements to cure speed or surface cure over that of the individual photoinitiators.

Many eutectic blends are also synergistic, and it has been shown that small amounts of multiple photoinitiators can be more effective than a high concentration of a single initiator.[49] Mixtures of the solid photoinitiators benzophenone, methylbenzophenone and trimethylbenzophenone will produce liquid eutectics such as Esacure TZM (80) and Esacure TZT (99). These eutectics also offer a small increase in cure speed from the sensitization effect via triplet energy transfer from the benzophenone to the methylbenzophenone (in TZM), and from methylbenzophenone to trimethylbenzophenone (in TZT), even though the wavelengths of absorption are relatively close (248 nm and 245 nm). A liquid eutectic blend of the solids benzophenone and Irgacure 184, Irgacure 500 (81), gives increased cure over 184 alone. In this case, the excited state of benzophenone, without an amine present, can break down peroxy radicals and improve the surface cure (Section 5.2). The mixture of benzophenone and Darocur 1173, Speedcure 210 (82), will provide similar advantages.

Solid mixtures, such as Irgacure 1300 (98), where a solid mixture of Irgacure 369 and Irgacure 651 produces more efficient cure in an ink than the individual photoinitiators, can also show synergy. Standard practice in formulation work often includes an intimate mixing/grinding step of the photoinitiators prior to dissolution in the binders. Intimate mixtures can sometimes produce "cage-like" products that are a little more efficient than the individual photoinitiators.

A combination of benzophenone and 4,4′-bis(dimethylamino)benzophenone, Michlers ketone (125) is an extremely efficient blend showing very high synergism. However, Michlers ketone has been shown to be carcinogenic and is no longer used. The diethylamino derivative, Ethyl Michlers ketone EMK (126), is less effective than MK and has the disadvantage of producing strong yellowing but is nevertheless an effective photoinitiator as well as being a very reactive hydrogen donor.

Other blends of photoinitiators are used to provide both surface and depth cure. This is most effective where the oxygen sensitivity of a phosphine oxide and its associated depth cure needs to be protected with some surface curing agent.[17,50] Darocur 4263 (89) comprises an 85/15 solution of Darocur 1173 and Lucirin TPO that gives rapid surface cure from the high concentration of short-wave absorption from 1173, with good depth cure from the low concentration of long-wave absorption of TPO. Similarly, Irgacure 1700 (86) provides a 75/25 mix of Darocur 1173 and Ciba BAPO-1. There are many combinations of hydroxyacetophenones and phosphine oxides available that are used more specifically for titanium dioxide white coatings.

Yet other mixtures are simply solutions of photoinitiators that are relatively insoluble or difficult to use with other liquid photoinitiators or with monomers such as GPTA and TPGDA, as in the Esacure KIP series (100–106).[51]

## 4.4    Migration and polymeric photoinitiators

### 4.4.1    Migration of low-molecular weight species

During the UV curing process, high concentrations of photoinitiators are often used to provide a high radical count and give good reactivity, and only part of this material is consumed. After full cure, this means that a significant amount of unchanged photoinitiator remains in the polymer.

Low-molecular weight additives in UV-curable formulations will be prone to migrate through a cured coating. These will include residual photoinitiators and their photoproducts, tertiary amines, unreacted low molecular weight monomers, UV absorbers, etc. Standard Soxhlet extraction of the cured coatings with various solvents have shown that photoinitiators, tertiary amines, and monomers such as ITX, BP, 1173, EDB, and HDDA can be readily extracted. The structures of the above materials are very different, so it is unlikely that migration is significantly influenced by the chemical structure of a small unit, and it is mainly the small size of the molecule that allows it to migrate within the coating and be extracted from the surface.[52] Contact of the cured ink surface by an absorbent material such as paper, card, or fatty substances has shown that these low molecular weight materials are readily transferred from the surface of the cured ink or coating to the contact material.

The majority of freezer food box packaging is now UV printed and many studies have found that traces of photoinitiators and residual monomers can be found absorbed in foodstuffs from these packages. The UK Food Standards Agency has found low levels of benzophenone in a multitude of foodstuffs from indirect food packaging, although these levels are well below the accepted limits and generally regarded as safe. Consequently, huge efforts have recently been directed into the development of non-migratable, polymeric species for this application.

Migration of photoinitiators from a coating or ink will depend on several factors,[53] including:

- The types of photoinitiators that are used and their photoproducts
- The foodstuff that is packaged and the type of packaging
- The ratio of the printed area to the weight of food packaged
- The cross-link density of the polymer
- The quality of the curing process, UV dose, etc.
- Any heat or pressure that may be applied in storage of the package, etc.

Many photoinitiators have a molecular weight of around 250 or less and this small molecular unit allows movement and migration of the photoinitiator through the UV-cured coating. The majority of commercial photoinitiators have molecular weights falling between 200 and 350 and all of them are quite likely to lead to some degree of migration similar to, if not as severe as, those shown by ITX, BP, EDB, and 1173.

It is not possible to put a "migration index" on each photoinitiator since migration is also influenced by other factors, not least of which is the polymer matrix itself. A restriction to migration is the entanglement of the free initiator molecule in the polymer and the degree of cross-linking will influence migration. Highly cross-linked hard polymers will reduce the extraction of these small molecules by entanglement, whereas a soft, loosely bound polymer will allow more freedom of movement of the initiator throughout the polymer. The migration of an initiator, just like its reactivity, depends on the nature of the medium in which it is used.

Printing of carton board or paper sheet usually entails "stacking" of the printed board for appropriate handling purposes. This allows the print surface to remain in contact with the lower surface of the upper board for a considerable time and under some pressure from the weight of the stack. The lower surface of the board may also be coated and the ink is then in contact with a partially absorbent film, depending on the coating (Figure 4.1). A result of stacking is "set-off" where low molecular weight materials are transferred from the surface of the ink to the underside of the printed board sitting above it.[54]

Migration under set-off conditions can occur

1. From the ink through the board to the lower surface
2. From the ink surface to the lower surface of the upper board
3. Through vapor phase transfer

Migratory materials absorbed in the lower surface in this way, with or without a polyolefin coating (pe), can then lead to contamination when a carton is formed.

When the carton is formed into shape, the lower surface of the board becomes the inside surface of the carton and any migratory materials

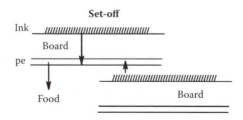

*Figure 4.1* Migration from a printed board into foodstuffs.

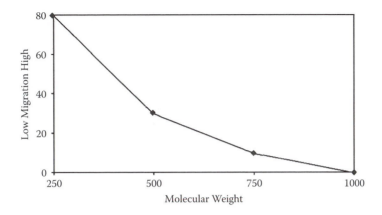

***Figure 4.2*** Migration vs. molecular weight.

absorbed on the lower surface will then be in contact with the contents of the carton. Any fatty substances or foodstuff contained in the carton is likely to "extract" these materials and contamination occurs. The problem is more severe in printed flexible polyolefin rolls that are used for wrapping foodstuffs.[55–57]

Photoproducts from the initiators can be controlled to some extent by modifying the structure of the initiator, and improvements will come with new initiators of a higher molecular weight polymeric nature, although Type I polymeric photoinitiators will always tend to produce low molecular weight species through the scission process.

Increasing the molecular weight of a photoinitiator is a crucial step in controlling migration. Photoinitiators with molecular weights around 200–250 will readily migrate. A much larger molecule will become trapped in the cured polymer network through physical entanglement and provide reduced migration. It has been shown that materials with a molecular weight of around 500 will lead to low levels of migration, and increasing the molecular weight toward 1000 will lead to practically zero migration. Figure 4.2, with an arbitrary migration scale from low to high, illustrates this point.

## 4.4.2   Polymeric photoinitiators

The commercial development of polymeric photoinitiators has been influenced by the regulatory process. Some countries exempt a product that is composed of a registered photoinitiator moiety attached to a polymeric structure, such as in S1 and S2 below, allowing much more cost effective development of polymerics. However, there are differing global definitions of a polymer and some reluctance to exempt anything in certain countries, so effective global marketing may still rely on a considerable set of toxicological tests for a new polymeric product. In Europe, this has not been a hindrance and the development of polymeric photoinitiators has been fast and furious.

Much academic work has been done on polymeric structures containing photoinitiator moeties and these can take on a variety of fashions using linear structures, pendant initiator species, initiator species within the chain itself, or dendritic structures. The majority of commercial polymeric initiators are based on the linear structures shown below where a photoinitiator species is attached by a linking group to one or both ends of a polymeric chain.[58–62]

S1.     **PI**–/\/\/\/\-polymer

S2.     **PI**–/\/\/\/\-polymer-/\/\/\/\–**PI**

This type of structure provides more cost effective production and produces a non-migratable molecule with a reasonable level of reactivity. These linear structures tend to give rise to viscous oils and resinous materials.

More recent structures (Figure 4.3) are based on pendant photoinitiators on a polymer chain and on three-dimensional multi-initiator groups in a simple dendrimeric structure. The latter type of structure can lead to less viscous materials and to solids that are readily soluble and behave more as individual photoinitiators despite their high molecular weight.

An advantage that may come with the use of polymeric photoinitiators is that an increase in cross-linking density may be gained. As one photoinitiator moiety produces a radical and is locked into the growing polymer chain, a second (or third) PI moiety may produce a second radical and chain-extend or cross link from its several photoinitiator functionalities. This may allow the formulator to reduce the amounts of multifunctional monomer to gain the same degree of cross linking and in turn leads to lower levels of monomer migration.

The reactivity of all of these polymeric materials is usually lower than that of the individual photoinitiator since the polymeric chain is in most

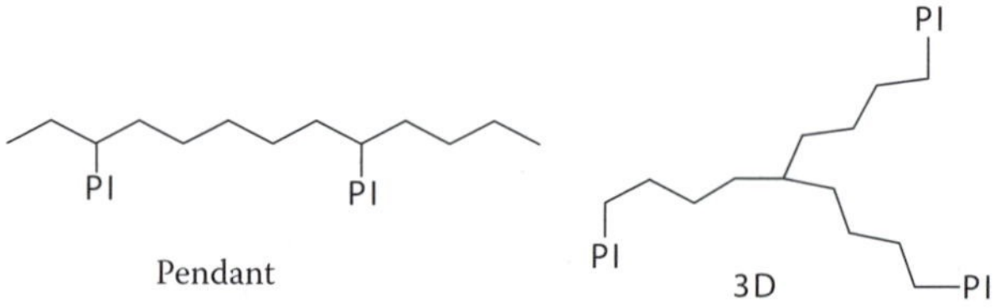

*Figure 4.3* Pendant and three-dimensional polymeric photoinitiators.

cases an unreactive "dead weight," but good cure speeds can be obtained, usually by adding a higher percentage of the polymeric species compared with the use of a standard photoinitiator.

There are three factors to consider when using these polymeric materials:

1. In the above structures the polymer chain, usually a polyether, is nonreactive, and the photoactivity of the molecule as a whole will depend on the molecular fraction of the initiator species.

   In Structure S1, above, the initiator fraction of the molecule may be as low as 30% of the molecular weight, and in Structure S2 the initiator fraction may be around 60%. Simple replacement of a standard photoinitiator in a formulation with the same concentration of one of these polymeric structures will lead to lower reactivity, and it may be necessary to increase the concentration of these polymeric initiators by 50% or more to achieve cure speeds comparable to those of a standard photoinitiator.

   For polymeric hydrogen donors such as the aminobenzoates, the polyether chain that is mostly used in these materials can also provide a source of hydrogen donor sites, and the overall reactivity is generally good despite the lower amine content. However, a problem arises with thin film curing where high amine content is required to counter oxygen inhibition and the polyether cannot contribute to this effect. For offset litho and flexo work where high amine content may be required, these polymeric amines cannot be regarded as simple replacements of aminobenzoates, but, with care, they can be used very effectively.

2. Polymeric photoinitiators have physical properties that are very different from those of standard photoinitiators. They are mostly highly viscous liquids and will affect the viscosity of the formulation, particularly if they are used in higher concentrations.

   Factors such as ink flow are affected and low viscosity formulations such as UV flexo inks and UV inkjet become more difficult to formulate using polymeric materials. Since the initiator is itself polymeric by nature, plasticizing effects may also affect the final properties of the ink. Additionally, the reactivity of any initiator or excited species depends to some degree on its mobility and its ability to interact chemically with other molecules. Larger polymeric species will be less mobile and the reactivity will be adversely affected.

3. If Type I photoinitiators are incorporated into a polymeric molecule, the scission process under UV will produce low molecular weight photoproducts that may themselves migrate.

An exception to this last point are the dimeric and polymeric hydroxyacetophenones (HAPS) such as Irgacure 127 (7) (dimeric), Esacure ONE (8) (dimeric) and Esacure KIP 150 (6). The alkyl radical that splits from these HAPs under UV forms acetone as a photo-product and acetone is sufficiently volatile to be driven off during the curing process.

For this reason, the majority of polymeric photoinitiators are based on Type II species such as benzophenone or thioxanthone that do not fragment. There are also a few high molecular weight photoinitiators that have been described as "low migratory," with molecular weights ranging from 340 to 800. These include Uvecryl P-36 (acrylated) (53), MW 428, which is a copolymerizable benzo-phenone with a low reactivity. Esacure 1001M (62) is a substituted benzophenone, MW 482, giving low migration. Irgacure 754 (74) is a bis-benzoylformate polymer for hard, clear coatings, MW ca 600. Irgacure 127 (7) is a very reactive bis-hydroxyacetophenone for clear coatings with a MW of 340. Esacure ONE (8) is a similar dimeric hydroxyacetophenone that will also give low migration.

### 4.4.2.1 Polymeric Type I photoinitiators

The Esacure KIP 150 series (6) of polymeric hydroxyacetophenone, absorb-ing at 262 nm, is used for very low odor, low migration, clear coatings. Irgacure PICS-1 (9) is a dendrimeric, high molecular weight, hydroxy-acetophenone that is under development. These will all release acetone on curing.

Omnipol 910 (15) is an alkylaminoacetophenone similar to Irgacure 369, absorbing at 325 nm, used for high speed, low migration inks. This will release an alkylamino radical that is very reactive and will be mostly involved in the curing process. Quadracure MMMP-3 (11) and BDMD-3 (13) are similar alkylaminoacetophenone derivatives with oligomeric ε-caprolactone substitution. These provide low odor and low extractables but will also release photoproducts from the primary scission.

### 4.4.2.2 Polymeric Type II photoinitiators

This category includes benzophenones such as Omnipol BP (54), Genopol BP-1 (55), Speedcure 7005 (56) and Goldcure 2700 (57). These will have no low molecular weight photoproducts but they still require a hydrogen donor that will, for food packaging applications, need to be polymeric itself.

Omnipol SZ is a polymeric sensitizer that can be used with Type II photoinitiators to improve cure.

Thioxanthones such as Genopol TX-1 (72), Omnipol TX (71) and Speedcure 7010 (70) also fall into this category. Again, a polymeric hydro-gen donor will be required in most cases.

### 4.4.2.3 Polymeric aminobenzoate hydrogen donors

Esacure A 198, Genopol AB1 (123), Genopol RCX02–766 (124), Omnipol ASA (120), Omnipol ASE (121) and Speedcure 7040 (119) are polymeric aminobenzoate hydrogen donors.

## 4.4.3 The "Nestlé list"

Following the appearance of "contaminants" in the contents of some food packages that were primarily linked with the migration of photoinitiators, the UV industry arrived at some consensus of regulation that would minimize future problems in this direction. Regulatory bodies such as the European Commission for Food Safety and Department of Environment, Food and Rural Affairs (DEFRA) in the UK agreed on risk assessments and asserted that they were concerned with "any materials" that could gain ingress into foodstuffs. A migration limit of 10 ppb when no toxicological evaluation has been made has been agreed on by the Resolution of the Council of Europe AP(2005)2. FDA legislation is somewhat different from the European directives and a variety of exposure scenarios have to be considered.

The Nestlé Initiative on Packaging Safety and Compliance[63] addresses the specific ink usage for decoration of packaging materials and defines the Nestlé Policy on Materials in Contact with Food (GI-31.008–1). Formulators declaring suitable compositions gain the Certificates of Compliance. The principle of low migration must be proven by the converter and comply with existing restrictions. Formulation must be specific for food packaging, and EU Directive 76/769/EEC relating to restrictions on the use of dangerous substances and Substances of Very High Concern (SVHC) must be taken into consideration. Restrictions on fanal pigments also apply and low molecular weight acrylate monomers should not be used. Lists of these restricted materials can be found in the guidance notes. Several UV ink formulations for low migration are recommended by Nestlé. The Nestlé authorized list of photoinitiators that may be used for low migration UV printing inks (Table 4.1) avoids those of questionable toxicity, odor, and high migration. The list is updated frequently.

## 4.5 Visible light curing

The medium-pressure mercury lamp emits a significant amount of visible light over 400 nm. Output lines at 404 nm, 436 nm, 546 nm, and 578 nm can be used by photoinitiators that absorb at the appropriate wavelength, although the two longer wavelengths are not used for general UV curing. In addition, a number of photoimaging techniques of a specialized nature use lasers at various discrete wavelengths in the visible range.

*Table 4.1* Nestlé List of Photoinitiators
for Low-Migration UV Inks

| Trade name | CAS Number | Mol. Wt. |
|---|---|---|
| Irgacure 369 | 119313–12–1 | 366 |
| Irgacure 379 | 119344–86–4 | 380 |
| Irgacure 2959 | 106797–53–9 | 224 |
| Irgacure 819 | 162881–26–7 | 418 |
| Irgacure 250 | 344562–80–7 | <1000 |
| Irgacure 127 | 474510–57–1 | 496 |
| Irgacure 754 | 211510–16–6 | 600 |
| Rhodorsil 2074 | 178233–72–2 | <1000 |
| Speedcure EHA | 21245–02–3 | 277 |
| Speedcure EDB | 10287–53–3 | 193 |
| Genocure PBZ | 2128–93–0 | 258 |
| Speedcure PDA | 223463–45–4 | >1000 |
| Speedcure CPTX | 142770–42–1 | 272 |
| Lucirin TPO-L | 84434–11–7 | 316 |
| Esacure 1001 | 272460–97–6 | 510 |
| Speedcure 7003 | | >1000 |
| Speedcure 7005 | 1003567–82–5 | 1196 |
| Speedcure 7006 | | >1000 |
| Speedcure 7010 | 1003567–83–6 | 1899 |
| Speedcure 7020 | | >1000 |
| Speedcure 7040 | 1003567–84–7 | 1066 |
| Speedcure BMS | 83846–85–9 | 304 |
| Polymeric PIs | | >1000 |

For general UV-curing applications, the thioxanthones and phosphine oxides utilize the 404 nm visible output for the curing of inks and titanium dioxide coatings, and camphorquinone will pick up the 436 nm line for dental applications.

Formulations that can be cured by visible light are susceptible to normal daylight operations and need to be handled under yellow safety lights. This prohibits their use to some degree and these visible light-sensitive formulations are not in common use. Applications are restricted to industrial light-sensitive areas specially designed for the application. A detailed assessment of "visible-light curing" is beyond the scope of this book, but a few systems are outlined below.

## 4.5.1 Titanocenes

Irgacure 784 (Titanocene)

Titanocenes absorb at 405 nm and 480 nm and are used for visible light curing. Irgacure 784 (42) is a fluorinated diaryl-bis-cyclopentadienyl titanium complex that is used with the argon ion laser at 488 nm for imaging applications. Irgacure 727 is a similar titanocene. The fluorine in the o-position gives chemical and thermal stability.

No primary radicals are formed and titanocenes are neither Type I nor Type II molecules. The excited state of the titanocene forms a complex with the acrylate monomer to produce a monomer radical that initiates polymerization. The titanocenes photobleach and can be used to produce very thick coatings. They are used mainly in high-resolution micro electronics and polyimide formulations.

## 4.5.2 Dibenzylidene ketones

A red shift in wavelength can be gained by extension of the ketone conjugation. The increased conjugation shown in dibenzylidene ketones produces a wavelength shift to above 400 nm, taking the absorption into the visible spectrum. Some of these highly conjugated species have been made with water-soluble end groups for use in photopolymers where high-power tungsten lamps emitting visible light are used. The example simply illustrates the type of structure that provides a red shift in absorption. This is a Type II system where the addition of an amine is required.

Substituted dibenzylidene ketone

## 4.5.3 1,2-Diketones

Apart from camphorquinone (75), several substituted benzils have been assessed, including phenanthrenequinone, but few are used commercially.

## 4.5.4   Ketocoumarins

Substituted ketocoumarins

5- or 7-Methoxy substituted ketocoumarins[64] absorb in the long-wave UV range up to 390 nm and are very good electron acceptors. Combined with aminobenzoates as electron donors, they provide efficient photoinitiating systems, but they are less effective when used with simple alkyl tertiary amines. 7-Dimethylamino substitution of the ketocoumarin red shifts the absorption into the visible range at 410–470 nm, making these systems more sensitive to argon lasers. N-phenylglycine is a more efficient hydrogen donor for these longer wavelength ketocoumarins.

## 4.5.5   Dye sensitized photoinitiation

Many dyes can be combined with tertiary amines or electron donors such as thiols to provide an initiating system that is sensitive to visible light.[65] Dyes can also be used as sensitizers with other photoinitiators. Most dyes are targeted to respond to specific laser outputs for use in photoimaging applications.

### 4.5.5.1   Dye plus coinitiator

Photoreducible dyes can act as electron acceptors from donors such as tertiary amines, N-phenylglycine, thiols, borate salts, etc. Acridinium dyes absorb around 460 nm. Xanthene (565 nm), thiazene (668 nm) and flavine dyes absorb at longer wavelengths. Mercaptans such as 2-mercapto benzimidazole (134) or 2-mercapto benzthiazole (133) are more efficient electron donors than the tertiary amines. These dye formulations may suffer from dark reactions of the very sensitive dye, which often leads to poor shelf life.

A commercial set of visible light photoinitiators has been developed by the Spectra Group, Ohio. These H-NU photoinitiators are based on the fluorone dye structure and various molecules absorb at 470 nm, 535 nm, and 635 nm. In combination with N-phenyl glycine or N,N-dimethyl-2,6-diisopropylaniline, the H-NU initiators can be activated by a 75-watt tungsten halogen lamp emitting around 585 nm. Cure to 6H pencil hardness can be achieved with 120 seconds of irradiation.[66,67]

### 4.5.5.2   Dye plus borate salt

Cyanine dyes absorb around 400–480 nm but the nitrogen cation on the dye makes them almost insoluble in UV binders due to their ionic nature. Triarylalkyl borates provide very efficient anions that can be combined with these dyes to give a good initiating system. Electron transfer from the borate anion to the dye leads to fragmentation of the anion and an alkyl radical is produced.

There are two major problems with these systems in that most borate salts are acid sensitive, leading to instability, and the systems are also extremely sensitive to oxygen inhibition, which means that normal UV cure in air is almost curtailed. Despite these problems, borate salts have been developed that can be efficiently used for visible-light curing systems.[68]

### 4.5.5.3   Dye plus HABIs

Hexaaryl-bisimidazolyls (HABIs) (38–41) are mainly used for visible light curing with appropriate sensitizers that absorb in the visible range. In particular, p-dimethylaminobenzylidene ketones are used as sensitizers, absorbing at 410–430 nm. However, HABIs are notoriously insoluble and are usually used in solvent-based formulations for specific thin film photoimaging.

### 4.5.5.4   Dye plus S-triazine

S-triazines with methoxy styryl substitution (28–30) absorb in the long wave UV range at 380 nm and in the visible range at 470 nm. They can be sensitized by many types of dye to produce an efficient photopolymerization. The problem of S-triazines is the generation of hydrogen chloride from the chlorine radical, producing an acid, which makes them unsuitable for some applications.

## 4.6   Water-based UV curing

Water-based UV formulations provide an entry into very low viscosity applications such as spray coatings for wood. More recent developments are now applying water-based UV technology on plastic substrates using Irgacure 754.[69] Developments in emulsion technology have led to UV-curable polyurethaneacrylate dispersions with molecular weights around 5,000–10,000 becoming commercial. Unsaturated polyester emulsions with molecular weights of 500–2,000 are also used. These typically contain 30–50% water that must be removed prior to UV cure. Hot-air or IR dryers are used to remove water, keeping the temperature below 60°C to avoid the loss of any photoinitiators that may be steam volatile.

At this point, prior to the UV-cure step, a tack-free finish can be achieved that enables the product to be handled if necessary. Irradiation from a standard UV source then leads to completion of the curing process, giving hard, scratch-free surfaces. No low molecular weight monomers are used in this type of formulation, which means there is no penetration of monomer into the wood that may lead to odor after cure. These dispersions have good matting properties. Three-dimensional UV arrays have been developed to eliminate shadow areas for curing furniture.

Water-soluble photoinitiators have been developed and marketed under the Quantacure label.[70–73] These were based on standard photoinitiator structures with side chains that contained water-soluble ionic groups such as quaternary ammonium salts or sulphonates. The quaternary salts were generally more reactive than the sulphonates, but under cure they sometimes produced aminelike odors. Their surfactant nature made them ideally suitable for micellar formulations, and excellent cure could be achieved in this context. However, the residual photoinitiator in the cured coating, with its ionic structure, tended to give the finished product a hydrophilic nature that led to poor long-term stability. Water-soluble photoinitiators with ionic structures have yet to find significant use and have become no longer commercially sustainable. There have been extensive reviews of water soluble photoinitiators.[74,75]

Standard oil-soluble photoinitiators can be used in water-based formulations but the problem of dissolution in the presence of water has to be overcome, particularly if solid photoinitiators are used.[76] Liquid photoinitiators provide much better dispersion characteristics and are easier to incorporate. Water-based emulsions of several standard oil-soluble photoinitiators are available such as Esacure DP 250 (92), SarCure SR 1126 (97), Esacure KIP EM (106), and Irgacure 819DW (107).

For clear coatings, Darocur 1173 (1) gives good reactivity, but is steam volatile to some degree and care is needed to avoid loss of photoinitiator at the water evaporation stage. Omnirad 102 (2) has similar reactivity, is less volatile, and can also be used in water-based formulations.

Irgacure 500 (81), a liquid eutectic mix of Irgacure 184 and BP, can lead to very good surface cure and is often combined with Lucirin TPO-L (23) for water-based use. An emulsion of the polymeric HAP, Esacure KIP EM (106), provides a non-volatile, short wave UV, water-compatible alternative for clear coatings.

If more depth cure is needed, small amounts of TPO-L (23) will give long wave UV absorption, better body cure, and adhesion. Alternatively, an emulsion of the bis-phosphine oxide, Irgacure 819 DW (107), can provide very fast, long wave depth cure in an aqueous environment. Typically, 1% TPO-L or 819DW is used with 2–3% Darocur 1173 or 2% Irgacure 2959.

Irgacure 2959 (4) is a short wave UV initiator with a non-ionic hydrophilic side chain that provides about 0.5% water solubility and good compatibility in water-based formulations, despite being a solid. Irgacure 2959 provides excellent cure and is often used in these water-based formulations for clear coatings.

Type II systems have also been shown to provide excellent cure, using benzophenone (50) and a simple amine, but aliphatic amines can give some odor and loss of gloss. A better alternative might be the liquid benzophenone mixture of Speedcure BEM (80) plus a low volatility liquid amine such as Speedcure DMB (114) or Speedcure EHA (116).

The polymeric benzoylformate, Irgacure 754 (74), can also be used in waterborne curing with polyurethane dispersions and gives very good cure speed and very low yellowing. Irgacure 754 has been used to provide a very thin water-based coating on difficult substrates such as plastics and metals, giving a UV base coat with much improved adhesion.[45]

Care must be taken with Type II systems to avoid low pH in water-based formulations, as protonation of the amine will occur and reduce the reactivity and some consideration of the effects of photoinitiators on the emulsifying agent may be needed.

Recent work has indicated that N-phenylglycine, NPG (128), may be a more efficient hydrogen donor than the aminobenzoates when used in an aqueous environment.[77] Radical ion separation in aqueous solution after electron transfer can reduce the reactivity of tertiary amines in a normal Type II electron/proton transfer mechanism. NPG produces radicals via a decarboxylation process that does not involve radical ions, and this reaction tends to be a more efficient process in aqueous medium. NPG does, however, suffer from some instability that can lead to poor shelf life.

## 4.7   Hydrogen donors

Activated hydrogen atoms are found on carbon atoms that are alpha to a hetero atom such as N, S, and O. These types of molecules can generally be used as hydrogen donors for Type II photoinitiators. Tertiary amines are regarded as very efficient hydrogen donors but thiols, alcohols, ethers, esters, etc., can also be used, though they are generally less effective.

If a Type II photoinitiator is used without a hydrogen donor in an oligomeric medium that contains available hydrogen atoms on the resin structure, then hydrogen abstraction can take place from the oligomeric backbone, promoting cross-linking and modification of the polymeric structure. This is one of the principles of photografting, such as modifying the surface properties of cellulose fibers, etc.

## 4.7.1   Tertiary amines

Tertiary amines are very efficient hydrogen donors that are widely used in the UV curing industry (see Table A.6).[78,79]

The reactivity of the radical species derived from the amine tends to be inversely proportional to the ionization potential of the amine (Section 3.7). Very weak bases produce very active radicals. Additionally, a hydroxy group in the amine structure also improves the reactivity; the family of "olamines" are more reactive donors than their "amine" counterparts. A secondary hydroxy group in the amine structure gives a slight improvement in performance yet again, over the primary "olamine."

Triethanolamine (113) is more reactive than triethylamine (110) and materials such as methyldiethanolamine, MDEA (112), have become standard simple amines for laboratory UV comparative and analytical work. The p-dimethylaminobenzoate esters, such as Speedcure EDB and EHA, are by far the most reactive of the several types of amine that are used in UV curing.

Reactivity of the types of amine increases:

$$(C_2H_5)_3N < CH_3N(C_2H_4OH)_2 < (CH_{\neg3})_2NCH_2CH(OH)CH_3$$

$$< (CH_3)_2NC_6H_4COOC_2H_5$$

Simple aliphatic amines are efficient, but their odor, volatility, and water solubility limit their use. These small molecules often migrate to the surface of the cured film, react with acidic materials in the atmosphere, and produce salts leading to a loss of gloss or a "bloom" in the film. Although these simple aliphatic amines such as MDEA (112) are widely used in academic studies, their use commercially applies mainly to the "cheap and cheerful" end of the market. Modified aliphatic amines such as dimethylaminoethyl benzoate, Speedcure DMB, are less migratory, give low bloom and low yellowing, but suffer from relatively poor reactivity.

Tertiary amines based on a deactivated aromatic ring produce radicals that are more reactive than those from simple alkyl amines, and esters of 4-dimethylaminobenzoic acid (115–124) are widely used. The aminobenzoate esters are also insoluble in water and can be used in the lithographic printing process where a water/ink balance needs to be carefully maintained and high cure speeds are required. As with photoinitiators, the concentration of the amine in a Type II application has a significant effect on the reactivity.[81] For example, using 3% Esacure 1001M (62) as the initiator with increasing levels of the aminobenzoate ester Speedcure EDB (115), a large increase in reactivity can be seen (Table 4.2).

*Table 4.2* Cure Speed vs. Amine Concentration

| | Cure speed m/min | | |
|---|---|---|---|
| % amine | 1.5 | 3 | 6 |
| Tack free | 35 | 50 | 80 |
| Full cure | 11 | 17 | 30 |

*Source:* Adapted from M. Cattaneo et al. *RadTech Eu. Conf. Proc.* 2003.

The alkylamino radical becomes one end group of the growing polymer chain. In Type II systems, a reduction in hardness of the coating compared with a Type I photoinitiator can be measured, showing that amines can have a small but definite plasticizing or softening effect on the finished product.

Acrylated amines and oligomeric amines are often used in Type II clear coatings to limit amine migration and reduce "bloom." These tend to be long-chain oligomers modified with diethylamine, dibutylamine, or ethanolamine and some have relatively low total amine content. Acrylated amines with a high amine content will tend to have a low acrylate functionality and are less likely to be "tied in" the polymer network. This may lead to higher migration, odor, and yellowing. The relationship between amine levels and acrylate functionality has been examined recently to provide low odor and migration.[80,82] Mobile amine species or unreacted material in the cured product generally lead to problems, and high molecular weight oligomeric amine acrylates are often the better choice. These types of amine synergists are used in fairly high concentrations of anything from 10%–50% to achieve reactivity in clear coatings. At these levels, the nature of the oligomeric amine, essentially a monomer, will affect both the rheology of the formulation and the final properties of the coating. There are, however, a huge number of these amine-modified resins on the market.

All amines will tend to yellow with time due to post-cure oxidative processes, particularly in outdoor applications, and the use of Type II photoinitiating formulations will always lead to some degree of long-term yellowing. In clear varnishes where clarity and water whiteness are required, Type II formulations represent the economical end of the market, and Type I photoinitiators, such as the hydroxyacetophenones, that do not require an amine, are more frequently used for higher-quality work. In UV inks where the colors may hide any yellowing, Type II systems are often preferred.

Tertiary amines are basic materials and, in low pH applications, such as acid resists and carboxyl modified resins, the acidic resin will neutralize the amine and inhibit photoactivity.

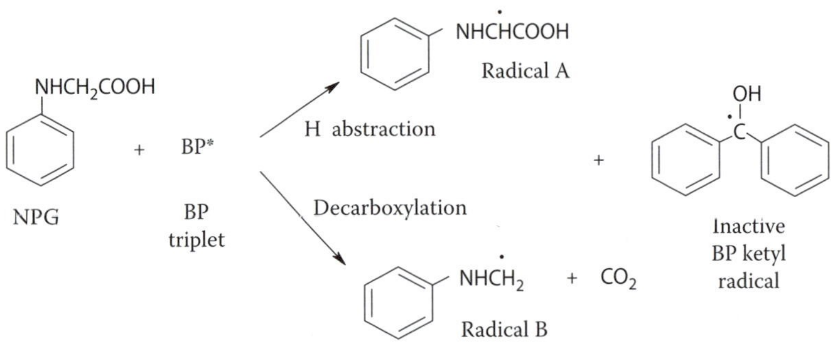

*Figure 4.4* Radical production from benzophenone triplet and N-phenylglycine, NPG.

## 4.7.2   *Alpha-amino acids*

α-Amino acids such as N-phenylglycine, NPG (128), are efficient hydrogen donors that are used in electronic and imaging applications. N-phenylglycine[77] can provide an active hydrogen atom from the α-carbon in a typical Type II tertiary amine abstraction process that is neutralized in the presence of acid. NPG does, however, preferentially produce radicals via a decarboxylation process, which does not rely on the basicity of the amine function. In this way, NPG can be used effectively as a hydrogen donor in acidic media where protonation of the more commonly used tertiary amine would occur. NPG works very efficiently in combination with ketocoumarins for visible light curing and as a coinitiator using reducible dyes in photoimaging.

A typical Type II abstraction process, shown in Figure 4.4, leads to radical A. A decarboxylation process, releasing carbon dioxide, preferentially leads to radical B. Both processes produce alkylamino radicals that are very reactive. In both cases, where benzophenone is used, the relatively unreactive benzophenone ketyl radical is also produced. A disadvantage of NPG is that the release of carbon dioxide can produce bubbles of gas in the coating. In addition, the decarboxylation is a facile process that can lead to instability in the formulation and poor shelf life.

Esters of N-phenylglycine are less reactive since they do not decarboxylate, but they can still provide reasonable levels of hydrogen donation and radical formation.

## 4.7.3   *Other types of hydrogen donor*

Thiols are also very efficient hydrogen donors but have the disadvantage of producing odor. There are a few commercially available thiols with relatively low, but still distinct, odor such as trimethylolpropane

$$-CH_2CH_2O- \longrightarrow -CH_2\overset{\bullet}{C}HO-$$

| Alkylether | Alkylether |
| donor | radical |

*Figure 4.5* Carbon-centered radical.

tris(3-mercaptopropionate) (132). These multifunctional thiols have been used to decrease the oxygen sensitivity in a system since thiols, like tertiary amines, are excellent oxygen scavengers. Thiols with acrylates will tend to react as a thiol-ene system, producing polysulphides, but one drawback of the thiol-ene reaction is the instability question and a poor shelf life may be the result. The odor disappears on cure as hydrogen donation takes place and the thiyl radical reacts to give a sulphide, but the prospect of using an inherently "smelly" formulation on the press would deter most printers. Hydrogen donors are only partially consumed during curing and the prospect of odor remains; it follows that thiols are seldom used commercially.

Alcohols, such as isopropanol, can be used but are much less efficient than tertiary amines. Secondary alcohols are more reactive than primary alcohols. Ethers and esters, although less efficient than amines, can also be used as hydrogen donors. They have no effect on oxygen inhibition, unlike amines, and are largely ineffective in thin films, but they can be used with some success in higher-weight screen inks. The carbon-centered radicals that are produced are not as reactive as alkylamino radicals (Figure 4.5).

Oligomeric resins containing ether, ester, and amide, etc., groups can act as hydrogen donors from the resin backbone. Type II photoinitiators used with these resins as donors will form very hard, high cross-linked coatings from what is essentially a photografting mechanism on the oligomer. These oligomers also find particular use in applications where there is low pH and where amines could not be used. Acid etch resists, which are used for the preparation of PCBs, often employ an anthraquinone photoinitiator that relies on the oligomeric ether structure of the resin for hydrogen donation.

## *References and further reading*

1. Fouassier, J. P., P. Jacques, D. J. Lougnot, T. Pilot. 1984. (ENSC Mulhouse) Lasers, photoinitiators and monomers: A fashionable formulation. *Polym. Photochem.* 5. 57–76.
2. Desobry, V. K., Dietliker, R. Huesler, H. Loeliger, W. Rutsch. 1988. *Polym. Paint. Color. J.* 178. 913.
3. Li Bassi, G., L. Cadona, F. Broggi. (Lamberti) Advances in low odor coatings: A new class of polymeric non-yellowing photoinitiators. *RadTech Eu. Conf. Proc.* 1987. 3–15 to 3–36.

4. Di Battista, P., G. Li Bassi, M. Cattaneo. (Lamberti) Selected applications of oligomeric hydroxyacetophenone photoinitiators. *RadTech NA, Conf. Proc.* 1990. 12–17.

5. Visconti, M., M. Cattaneo, G. Li Bassi. (Lamberti) Esacure KIP 150, a non-migrating, non-benzaldehyde releasing photoinitiator. *RadTech NA, Conf. Proc.* 1998. 28–30.

6. Kohler, M. (Ciba). 1997. A versatile alpha-hydroxyketone photoinitiator. *Eur. Coat. J.* 12, 1118–1120.

7. Fuchs, A., T. Balle, S. Ilg, R. Husler. (Ciba) New raw materials for UV inks and varnishes. *RadTech Eu. Conf. Proc.* 2003. 507–511.

8. Villeneuve, S., K. Studer, J-L. Birbaum, E. Brendle, S. Ilg. (Ciba) New raw materials for low emission coatings and inks formulation. *RadTech Eu. Conf. Proc.* 2007. Graphic Arts.

9. Li Bassi, G., G. Norcicni, S. Romagnano, M. Visconti. (Lamberti). PCT Int. 2005. Patent Appln. WO 014,515.

10. Rutsch, W., G. Bover, R. Kirchmayer. (Ciba) New photoinitiators for pigmented systems. *RadTech NA, Conf. Proc.* 1984. 5–49 to 5–71.

11. Misev, L. (Ciba) Selection criteria for high performance photoinitiators in pigmented systems. *RadTech Asia, Conf. Proc.* 1991. 404–410.

12. Wang, J. R. C., R. Yang, L. Yang, C. Chiu. (Chitec Technology Co. Ltd.) New liquid photoinitiators for light and dark pigmented ink systems. *RadTech NA. e/5*, 2004.

13. Misev, L., V. Desobry, K. Dietliker, R. Husler, M. Rembold, G. Rist, W. Rutsch. (Ciba) A novel photoinitiator for modern technology. *RadTech Eu. Conf. Proc.* 1989. 359–367.

14. Chang, C. H., D. Wostratzky, A. Mar. A novel photoinitiator for radiation curable coatings. *RadTech NA, Conf. Proc.* 1990. 1–11.

15. Berner, G. R., Kirchmayer, G. Rist. 1978. *J. Oil Col. Chem. Assoc.* 61, 105.

16. Christensen, J., A. F. Jacobine, C. J. V. Scianio. 1981. *J. Radiat. Curing* 8, 12.

17. Bankowski,H., E. Beck, M. Lockai, R. Noe, W. Reich, C. Glotfelter, K. Sass. (BASF) Acylphospine oxides as photoinitiators. *RadTech NA, Conf. Proc.* 2000. 545–559.

18. Baxter, J. E., R. S. Davidson, H. J. Hageman, K. A. McLauchlen, D. G. Stevens. 1987. The photo-induced cleavage of acylphosphine oxides. *J. Chem. Soc. Chem Commun.* 73.

19. See 17.

20. Cui, H., R. Nagarajan. (ChemFirst) Photoinitiators for curing white coatings. *RadTech NA, Conf. Proc.* 2000, 375–385.

21. Dietliker, K., G. Hug, R. Kaeser, M. Kohler, U. Kolezak, D. Leppard, L. Misev, G. Rist. W. Rutsch. Novel, high performance bisacylphosphine oxide photoinitiators, (BAPO-1). *RadTech NA. Conf. Proc.* 1994. 693–707.

22. McGiniss, V. D. 1975. *J. Radiat. Curing* 2(1), 3.

23. Coyle, J. D. 1987. Eur. Pat. Appl. EP 0271, 195.

24. Kura, H., J. Tanabe, H. Oka, K. Kunimoto, A. Matsumato, M. Ohwa. New oxime ester photoinitiators for color filter resists. *RadTech Report*, May/June 2004, 30–35.

25. Bradley, G., R. S. Davidson. 1995. *Rec. Trav. Chim.* 114, 528.

26. Photoacid generators for microlithography. Ciba Specialty Chemicals. Data sheet. 2006. www.ciba.com/cgi725_V202.pdf.

27. Visconti, M., M. Cattaneo, A Casiraghi, G. Norsini, G. Li Bassi. (Lamberti) X. Allonas, J. P. Fouassier. LFC 1001, a novel, low odor photoinitiator for dark, pigmented systems. *RadTech NA, Conf. Proc.* 2000. 414–426.

28. Visconti, M., M. Cattaneo. (Lamberti) Difunctional photoinitiators. *RadTech Eu. Conf. Proc.* 2003. 207–212.

29. Dietliker, K. (Ciba). A novel photolatent base catalyst for UV-A clearcoat applications. *RadTech Eu. Conf. Proc.* 2005. 471–478.

30. Dogan, N., H. Klinkenberg, L. Reinerie, D. Ruigrok, P. Wijmands. (Akzo) K. Dietliker, et al. 2006. Fast UV-A Clearcoat. *RadTech Report* March/April, 43–52.

31. Jung, T., K. Dietliker, J, Benkhoff. (Ciba) New latent amines for the coatings industry. www.cibasc.com.

32. Dietliker, K. (Ciba). Advancements in photolatent amines: Expanding the scope of photolatent base technology. *RadTech NA. Conf. Proc.* 2008.

33. Ledwith, A., J. A. Bosley, M. D. Pubrick. 1978. *J. Oil Color Chem. Assoc.* 61, 95.

34. Ruhlmann, D., J. P. Fouassier. Relations structure—proprietes dans les photoamorceurs de polymerisation—1. Derives de benzophenone. *Eur. Polym. J.* 27 (1991), 991.

35. Fouassier, J. P., D. J. Lougnot. (ENSC Mulhouse) 1990. Laser spectroscopy of C-S bonded photoinitiators: BMS, an aryl aryl sulphide derivative. *Polym. Commun.* 31, November, 418–421.

36. Green, W. A. (Great Lakes Fine Chemicals) Sulphur activated benzophenone photoinitiators. *RadTech Asia. Conf. Proc.* 1995. 271–281.

37. Anderson, D. G., R. S. Davidson, J. Elvez. 1995. An appraisal of 2,4-diethylthioxanthone as a photoinitiator. *Surf. Coat. Int.* 11, 482.

38. Green, W. A., P. N. Green, A. W. Timms. (IBIS) Propoxy substituted thioxanthones as photoinitiators and sensitizers for the UV curing of pigmented systems. *RadTech Eu. Conf. Proc.* 1991, Paper 51.

39. Green, W. A., A. W. Timms. (IBIS) CPTX, a novel photoinitiator and sensitizer for the UV curing of pigmented systems. *RadTech NA. Conf. Proc.* 1992, 33–37.

40. Allen, N. S., N. Sallah, M. Edge, T. Corralles, M. Shah, F. Catalina, A. Green. (Manchester Met. Univ.). 1997. Photochemistry and photoinitiation properties of novel 1-chloro substituted thioxanthones—II. Influence of 4-oxo and 1-phenylthio substitution. *Eur. Polym. J.* 33(10–12) 1639–1643.

41. Allen, N. S., M. Edge, F. Catalina, C. Peinado, A. Green. (Manchester Met. Univ.). 1999. New trends and developments in the photochemistry of thioxanthone initiators. *Trends Photochem. Photobiol.* 5, 7–16.

42. Meier, K., H. Zweifel. (Ciba). 1986. Thioxanthone ester derivatives: Efficient triplet sensitizers for photopolymer applications. *J. Photochem.* 35, 353–366.

43. Hulme, B. E., J. J. Marron. 1984. *Paint Resin* 54, 31.

44. Hu, S., D. C. Neckers, R. Popielarz. 1998. Florescence probe technique (FPT) for measuring the relative efficiencies of free radical photoinitiators. *Macromolecules* 31, 4107–4113.

45. Fuchs, A., T. Bolle, S. Ilg, R. Huesler. (Ciba). New generation of photoinitiators for UV inks and varnishes. *RadTech NA. Conf. Proc.* 2004. 14.4.

46. Biry, S., A. Naisby, R. Telesca, G. Sitta, A. Di Matteo. (Ciba) Advanced photo-initiator solutions for UV ink jet applications. *RadTech Eu. Conf. Proc.* 2007. Ink jet/digital.

47. Adamczak, E., L. A. Linden, J. F. Rabek, A. Wrzyszczynski. Camphorquinone photocuring of dental material. *RadTech Asia. Conf. Proc.* 1995. 196–203.

48. Scott, T. F., W. D. Cook, J. S. Forsyth. 2002. Photo-DSC cure kinetics of vinyl ester resins. *Polymer* 43, 5839–5845. Review.

49. Bishop, T. E. (DSM Desotech Inc.) Multiple photoinitiators for improved performance. *RadTech NA. Conf. Proc.* 2008.

50. Cattaneo, M., M. Visconti. (Lamberti) Liquid mixtures of photoinitiators based on MAPO. *RadTech NA, Conf. Proc.* 1998. 731–734.

51. Di Battista, P., M. Cattaneo, G. Li Bassi. (Lamberti) New optimised oligomeric hydroxyacetophenone photoinitiators. *RadTech Asia, Conf. Proc.* 1991. 398–403.

52. Beck, E., M. Lockai, E. Keil, H. Nissler. (BASF) Studies into the emissions from radiation cured coatings. *RadTech Eu. Conf. Proc.* 1995. 351–361.

53. Studer, K. (Ciba). Reduction of photoinitiator migration by control and optimisation of the UV-curing process. *RadTech NA. Conf. Proc.* 2008.

54. Ruter, M. (FABES). UV printing inks in food contact materials—migration and set-off problems. *RadTech Eu. Conf. Proc.* 2005. Vol.2. 137–146

55. Lin, A. (Sovereign Spec. Chem.). FDA Compliant Testing for UV curable materials used in flexible packaging. *RadTech Eu. Conf. Proc.* 2003. 325–335.

56. Veraart, J. R. (Keller & Heckman). How to demonstrate compliance with food packaging applications. *RadTech Eu. Conf. Proc.* 2007. HSE.

57. Benz, H., M. Hoburger, B. Pettinger, M. Ruter. (FABES) UV-printed food packages—the legal and technical status 2 years after the ITX crisis. *RadTech Eu. Conf. Proc.* 2007. Printing, varnishing and laminating for the packaging industry.

58. Anderson, D. G., N. R. Cullum. (Lambson Ltd.) R. S. Davidson. UV curing properties of polyalkyleneglycol modified amine synergists and ketonic photoinitiators. *RadTech Eu. Conf. Proc.* 1997. 282–290.

59. Davidson, R. S. (Citifluor Ltd.) R. Burrows, D. Illsley. Proc. PRA Materials and Markets Conference. Multifunctional photoinitiators (MFPIs)—a new concept. Bredbury (UK), 2002. Paper 4.

60. Herlihy, S. (Coates Lorilleux) The use of multifunctional photoinitiators to achieve low migration in UV cured printing. *RadTech NA. Conf. Proc.* 2002. 413–427.

61. Bertens, F. (IGM Resins) New polymeric photoinitiators for graphic arts. *RadTech Eu. Conf. Proc.* 2005. 473–478.

62. Anderson, D. G., C. A. Bell (Lambson Ltd.) R. S. Davidson, P. Sellars. Mono and Bis substituted polymeric aminobenzoates as amine synergists for UV curing. *RadTech Eu. Conf. Proc.* 2005. 437–443.

63. Simian, H. Packaging food safety at Nestlé. *RadTech Europe 2009 Conf. Proc.*

64. Williams, J. L. R., D. P. Specht, S. Farid. 1983. *Polym. Eng. Sci.* 23, 1022.

65. Neuenfeld, S., H-J. Timpe. (Merck) Dyes in photoinitiating systems. *RadTech Eu. Conf. Proc.* 1991. Paper 64, 798–802.

66. Neckers, D. C., R. F. Wright, J. Shi, D. Martin. (Spectra Group Ltd.) New visible initiator systems with panchromatic sensitivity. *RadTech NA. Conf. Proc.* 1992. 28–32.

67. Marino, T. L., D. Martin, D. C. Neckers. (Spectra Group Ltd.) Novel fluorone visible light photoinitiators. *RadTech NA. Conf. Proc.* 1994. 169–179.

68. Cunningham, A., M. Kunz. (Ciba) Acid-stable dye-borate electron transfer photoinitiators. *RadTech NA, Conf. Proc.* 1998. 38–41.

69. Weikard, J., E. Luhmann, S. Sommer. (Bayer AG) The new frontier: Waterborne UV coatings for plastics. *RadTech Eu. Conf. Proc.* 2007. Application Properties.

70. Allen, N. S., F. Catalina, P. N. Green, W. A. Green. (Manchester Met. Univ.). 1986. Photochemistry of thioxanthones—II. A spectroscopic and flash photolysis study on water soluble structures. *Eur. Polym. J.* 22(5), 347–350.

71. Allen, N. S., F. Catalina, P. N. Green, W. A. Green. 1986. (Manchester Met. Univ.) Photochemistry of thioxanthones—V. A polymerization, spectroscopic and flash photolysis study on novel, water soluble, methyl substituted thioxanthones. *Eur. Polym. J.* 22(11), 871–875.

72. Allen, N. S., F. Catalina, J. L. Mateo, R. Sastre, W. Chen, P. N. Green, W. A. Green. (Manchester Met. Univ.). 1989. Photochemistry and photopolymerization activity of water soluble benzophenone initiators. ACS. *Polym. Mat.: Sci. & Eng.* 60, 10–14.

73. Green, P. N., W. A. Green. (Int. Bio-Synthetics Ltd.) Novel, water soluble, copolymerizable benzophenone photoinitiators. *RadTech Eu. Conf. Proc.* 1989. 383–391.G.

74. Green, W. A. (Great Lakes Fine Chem. Ltd.). 1994. Water soluble photoinitiators: A Review. *Polym. Paint. Color. J.* 184(4358), 474–477. Also in *Eur. Coat. J.* 5/1994, 274–291.

75. Green, W. A., A. W. Timms. in *Radiation Curing in Polymer Science and Technology. Vol. II. Photoinitiator Systems.* J. P. Fouassier and J. F. Rabek, Eds. Amsterdam: Elsevier Science. Chapter 7, 375–434. (Review).

76. Wehner, J. Ohngemach. (Merck) Photoinitiator efficiency in water borne systems. *RadTech Asia, Conf. Proc.* 1991. 423–426.

77. Uhlrich, G. (Vienna Univ.) Investigation of phenylglycine derivatives as coinitiators. *RadTech Eu. Conf. Proc.* 2005. Vol. 2. 417–422.

78. Allen, N. S., L. K. Lo. (Manchester Met. Univ.) M. S. Salim. (Harcros) Applications and properties of amine synergists in UV and EB curable coatings. *RadTech Eu. Conf. Proc.* 1989. 253–268.

79. Visuranathen, K., C. E. Hoyle, S. E. Jonsson, C. Nason, K. Lindgren. (Univ. Southern Mississippi). 2002. Effect of amine structure on photoreduction of hydrogen abstraction initiators. *Macromolecule* 35(21), 7963–7967.

80. Kelly, P., A. Little, C. Bell. (SI Group UK) Synergy or enemy—A basic dilemma. *RadTech EU Conf. Proc.* 2007. Application Properties.

81. Cattaneo, M., R. Filpa, P. Soragna, M. Visconti. (Lamberti) Use of a novel bifunctional photoinitiator in graphic arts. *RadTech Eu. Conf. Proc.* 2003.

82. Oliver, J. (Sartomer) The use of amine synergists for low odor and low migration applications. *RadTech Eu.Conf. Proc.* 2007. Graphic Arts.

# chapter five

# Factors affecting the use of photoinitiators

At one time, the use of photoinitiators was considered to be a bit of a "black art." Some photoinitiators would work well whereas others did not seem to work at all for a particular application, and it was soon realized that factors such as the use of pigments, film thickness, viscosity, etc., all had an influence on the performance of a photoinitiator. Many factors determine which photoinitiators should be chosen for an application and how the different types are used. A few basic principles are outlined here.

## 5.1 Matching the photoinitiator absorption to the UV source

### 5.1.1 Absorption properties

The ultraviolet spectrum most commonly used for UV curing covers wavelengths from 200 nm to 400 nm. In addition, the UV-VIS spectrum from 400 nm to 450 nm is also important for pigmented media. The medium-pressure mercury lamp is the most common source of UV light industrially. The MPM lamp provides a series of line outputs, although there is a fairly good wavelength distribution throughout the UV spectrum (see Chapter 1).

The absorption pattern of the thioxanthone CPTX (69), shown against the background of the output of a medium-pressure mercury lamp in Figure 5.1, indicates a low absorption at 350–420 nm that will pick up the 366 nm and 404 nm UV lines and gives good depth cure when used in low concentrations.

Photoinitiators that have a strong absorption in the short wave UV shown at 240–270 nm will pick up the 254 nm output line and give good surface cure. The 313 nm output will provide some depth cure as well as surface cure, but is normally associated with high-speed photoinitiators. If some of the light output is screened by other materials such as pigments, it is important to match the concentration of photoinitiator to the absorption profile at the wavelength of the available light. For example, Irgacure 369 (12) has a strong absorption at 320 nm tailing off to 410 nm.

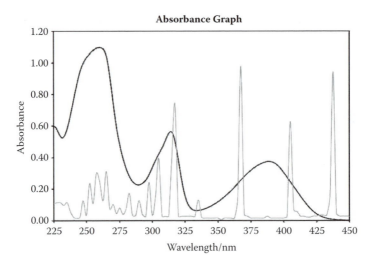

*Figure 5.1* Typical absorption pattern of a thioxanthone, CPTX.

If the 313 nm output line is not screened by pigments and can be used, 0.25% concentration will give maximum cure for a screen ink. If this line is masked by a high concentration of pigment and only the 366 nm or 404 nm lines are available (where absorption is much lower), concentrations up to 3.0% may be needed for maximum cure.

Although it is customary to match the absorption profile of the photoinitiators that are used in a formulation to the UV lamp output,[1,2] most commercial photoinitiators will respond well to the MPM lamp and this matching factor is of only minor importance. More care is needed if alternative UV sources are used with more specific outputs such as doped UV lamps, or LED UV light, which has a single output at 365 nm or 395 nm. Matching photoinitiator absorption to the UV wavelength then becomes more important.

## 5.1.2   Formulating for LED UV

LED lamp systems are now being used for UV inkjets where lightweight, "mobile" lamps need to follow the ink drops and provide rapid cure. Lamp systems have also been developed to provide an instant "fix" of the droplet, followed by a second lamp of greater intensity to give full cure.[3] Formulating for LED UV lamp systems requires photoinitiators that absorb most efficiently in the long wave UV range at 365 nm and 395 nm, since these are the only output lines of most SLMs or "blocks" of LED lamps. Cure will depend on the coating thickness and the effectiveness of the initiator(s) in providing both surface and depth cure. For inkjet in particular, surface cure is a major problem due to the very low viscosity of the formulation, which allows oxygen to penetrate rapidly and inhibit cure. In standard UV curing, the different wavelengths are used to provide full cure, with both surface and depth cure being promoted by short wave and long wave photoinitiators respectively. For LED systems with a single wavelength output at 395 nm, this arrangement is not possible, and formulating is very difficult. Where a second LED can be used, emitting at 365 nm, formulation becomes less of a problem.

### 5.1.2.1   Type I photoinitiator LED systems

The phosphine oxides Lucirin TPO, TPO-L and Irgacure 819 (22–24) all absorb at 395 nm and will provide good depth cure with an LED array, particularly since they photobleach and allow light to penetrate to depth. Phosphine oxides, when cured in air, are very sensitive to oxygen inhibition and will not give surface cure, leaving a tacky surface. In standard UV systems without nitrogen blanketing, this is overcome by using Darocur 1173 (1) or Irgacure 184 (3) to provide surface cure at 254 nm in combination with the phosphine oxides. In this particular case, neither 1173 nor 184 will respond to LED UV at 395 nm, and surface cure is very difficult to achieve. Under nitrogen, phosphine oxides will be much more effective.

Few other initiators can be used at 395 nm with LEDS to provide surface cure in a Type I system. Irgacure 369 (12) and 379 (14) absorb at 320 nm but have a "tail" that absorbs up to 400 nm, although the absorption is quite weak at this wavelength. While Irgacure 369 or 379 should respond to a 395 nm LED and provide some degree of surface cure, adding a 365 nm LED to this combination of photoinitiators would be much more effective. Irgacure 907 also has a "tail" in its absorption that may just reach 395 nm, and again, this initiator will respond much better to a twin LED lamp system comprising both outputs.

It is possible to use reasonable concentrations of tertiary amines with the phosphine oxides, which will counter oxygen inhibition at the surface, but phosphine oxides are sensitive to nucleophilic attack and materials such as ethanolamines will rapidly destroy them. The aminobenzoates such as Speedcure EDB (115) and EHA (116) have a lower basicity and will be more suitable than aliphatic or oligomeric amines, but care is needed and the shelf life of the formulation may be affected.

Thiols and polysulphides are also effective oxygen scavengers and may be more effective for enhancing surface cure without resorting to the use of tertiary amines or to using nitrogen blanketing, but with these thiols there is always the problem of odor.

### 5.1.2.2   Type II photoinitiator LED systems

The thioxanthones, such as ITX (63) and DETX (65), absorb at 395 nm and would be suitable for LED formulations combined with amines such as aminobenzoates. Surface cure may be less of a problem since an amine will be present, but there are still no other initiators that can be used for direct UV absorption at 395 nm to provide surface cure. Higher concentrations of thioxanthones to give good surface cure would simply limit depth cure by reducing the light penetration. As with the Type I initiators, adding a second LED emitting at 365 nm makes formulating somewhat easier. The substituted benzophenone (BP) Speedcure BMS is a fast initiator that has some absorption at 365 nm, although it is not great at that wavelength, and would give some surface cure.

### 5.1.2.3   *The use of ITX as a sensitizer for LEDs*

Long wave sensitization by thioxanthones will probably be the most effective formulating strategy for LEDs. Sensitization of Irgacure 907 (10) and Irgacure 369 (12) is well documented (see Section 5.4) and can lead to very high cure speeds. For example, adding 0.5% ITX to 2% Irgacure 907 will more than double the cure speed, and a similar addition to Irgacure 369 can triple the cure speed. In both these cases, the light absorption is through the long wave ITX followed by energy transfer. Using 3% Irgacure 907 plus 1% ITX in a 395 nm LED system for a UV ink or coating would lead to good cure since the ITX will sensitize the high concentration of 907 without any light being absorbed by the 907. Faster cure would result from the use of Irgacure 369 plus ITX, and the addition of an amine to this mixture would help with surface cure. As before, the use of twin LEDs at 365 nm and 395 nm would give improved surface cure.

Sensitization of Type II initiators could also be effective since ITX will sensitize the substituted BPs Speedcure PBZ (58) and Speedcure BMS (60) to some degree. Coupled with an amine, the (sensitized) PBZ will provide surface cure and the ITX (395 nm absorption) will provide depth cure. Similarly, a reasonably high loading of BMS (plus amine) will give much greater reactivity coupled with 1–2% ITX. Sensitization effects are, however, much more efficient with the ITX/907 or 369 combinations.

## 5.2   Oxygen inhibition

### 5.2.1   *Radical–oxygen interaction*

Free radical chemistry suffers from the effects of oxygen in the coating and at the surface of the print during cure.[4,5] This happens in two ways:

First, oxygen will quench the excited state of a photoinitiator and prevent the formation of radicals.

$$(PI) \text{ triplet} + O_2 \rightarrow (PI) \text{ ground state} + O_2{}^* \rightarrow O_2 + \text{heat}$$

This is more likely to happen with Type II photoinitiators that have longer triplet lifetimes. Type I photoinitiators with very short lifetimes are less likely to be affected by energy quenching.

Second, oxygen can scavenge radicals, reducing the radical count. Oxygen will react with the free radicals to produce peroxy radicals that have a low reactivity toward the acrylate function, effectively removing radicals from the system. Radicals from both the initiator and from the active growing polymer chain can be affected.

$$R \cdot \ + O_2 \longrightarrow ROO \cdot$$

Active                              Inactive

Peroxy radicals

The peroxy radicals are relatively stable but may terminate the growing chain by radical–radical interaction. This results in a low molecular weight polymer. Any oxygen in the coating or absorbed from the air at the surface during cure will reduce the rate of polymerization by scavenging radicals.

Thicker coatings are less likely to suffer from this effect, since the oxygen in the depth of the coating is rapidly depleted and only surface cure may be affected. Thin coatings and inks, where oxygen from the air can be absorbed to the full depth of the film, are likely to suffer most. Viscosity also has a major part to play in oxygen uptake and transfer during cure.

Practical measures that can be used to offset oxygen inhibition include:

1. Nitrogen or carbon dioxide inertion to reduce the oxygen level at the film surface
2. The incorporation of small amounts of wax in the coating to provide a surface barrier
3. The use of excess tertiary amine to scavenge peroxy radicals
4. High photoinitiator concentration and high intensity light source
5. The use of short wave UV to seal the surface prior to full cure.

A high concentration of photoinitiator, particularly absorbing in the short wave UV, will produce a high radical count and take up oxygen, leading to faster cure, particularly at the surface.

Figure 5.2 illustrates the significantly reduced rate of conversion in air compared to that under nitrogen. In a pure oxygen environment, radicals are mopped up so efficiently that cure is almost impossible.[6]

## 5.2.2   Nitrogen and carbon dioxide inertion

Nitrogen blanketing, and more recently the use of carbon dioxide, is a difficult mechanical problem associated with the press design, used mainly

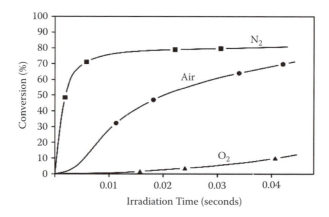

*Figure* 5.2 The influence of oxygen on the conversion of a diacrylate resin. (C. Decker. *RadTech Eu. Conf. Proc.* 1987. Reprinted with permission.)

on web, and its use depends very much on the type of print that is run. The cost of the gas and the flow rates can be a difficult problem to assess for effective use. Nitrogen blanketing can improve both the surface cure (scratch test) and the depth cure (thumb twist test). It is not necessary to attain zero oxygen concentration at the surface. Levels of around 0.5% oxygen will provide more than adequate blanketing in most cases. This will often allow the photoinitiator concentration to be reduced, still attaining the required cure speed, and will lead to better cross linking and fewer termination reactions from a lower radical count and lower migration. For example, in the printing of furniture foils, a high cross-link density is required for a 10-micron clear coating that gives good scratch resistance.[7] To obtain full cure giving 30 acetone double rubs, a maximum cure speed is around 80 m/min. Curing at higher speeds leads to softer surface cure and some odor from uncured monomers. Under nitrogen, the photoinitiator concentration can be reduced very significantly from 5% to 0.5%, still retaining full cure. These lower levels of photoinitiator bring lower migration and provide more stable, uniform, and predictable surface cure, which in turn brings better release factors for labeling. Similar criteria apply to amine synergists, which can, under nitrogen, be used at much lower levels.

Temperature has a greater influence under nitrogen and higher cure temperatures will bring better and more consistent cure rates. Where high-viscosity coatings are used, pre-heating the formulation to 40°C will give more consistent coating and improve both cure and adhesion. One downside of nitrogen blanketing is that the gas flow can take up volatile materials that may fog up the lamp and the quartz windows.

Retaining nitrogen flow to maintain low oxygen levels can result in the use of high levels of gas. Gas seal kits for UV units are available that will reduce the usage.[8] These are fitted close to the substrate before and after the lamp, and oxygen levels as low as 30 ppm can be achieved at speeds up to 300 m/min. Acrylate conversion levels under nitrogen rise from 75% in air to 96% or, alternatively, to achieve an expected 75% conversion, cure speeds can be increased from 20 m/min to 56 m/min. Another possibility under nitrogen is that lamp power can be reduced to achieve the same cure.

Carbon dioxide behaves similarly to nitrogen for inertion and there has been much recent work done on the use of carbon dioxide as an inert gas for the curing of three-dimensional objects.[9] Carbon dioxide is 1.5 times more dense than nitrogen or air and will collect or be contained in an open-top vessel. This factor is well appreciated in the chemical industry, where reactors and containers require servicing and oxygen levels have to be monitored before any access to such vessels is allowed. Cladding the walls of an open-top vessel with aluminum will provide reflection from a UV lamp placed in the opening. Any 3D object placed in the vessel will then be completely illuminated by multiple reflections. Filling such a vessel with carbon dioxide takes very little gas and only a very small gas flow is needed. There is some gas turbulence due to heat from the lamps, but oxygen levels around 2% can easily be achieved and this provides significant advantages in cure. Since surface cure is improved dramatically under low oxygen levels, there is less need for short wave UV and UV B and C becomes less important. In this respect, the less harmful UV A lamps may be all that is required for full cure.

## 5.2.3  The effect of viscosity on cure

Viscosity has a large influence on the rate of transfer of oxygen into a film and the corresponding inhibition of cure. Low-viscosity printing inks such as gravure, flexo, and particularly inkjet, are very susceptible to oxygen diffusion into the ink and the resulting inhibition leads to poor cure. For very low-viscosity formulations such as UV inkjet, oxygen transfer into the film is very rapid and becomes a severe inhibiting factor, even at very low temperatures where cure speeds are reduced. More viscous, buttery offset inks and thicker screen inks are less likely to suffer from inhibition, and nitrogen blanketing is less effective for these inks.

In air, the cure rate drops rapidly at lower viscosities as oxygen transfer into the coating is facilitated and inhibition increases. Under nitrogen, the opposite occurs and the cure rate increases more significantly with decreasing viscosity and increased mobility of species.

*TABLE 5.1* Cure Rate Inhibition of Various Types
of Ink in Air

| Type of ink | Viscosity poise | Thickness microns | Inhibition % |
|---|---|---|---|
| Ink-jet | 0.05–0.1 | 8–10 | 70–80 |
| Gravure | 1–3 | 4–7 | 50–60 |
| Flexo | 3–12 | 2–3 | 40–50 |
| Offset | 200–500 | 1–2 | 20–30 |

An indication of the inhibition of various types of inks in air, compared with their reactivity under nitrogen, are shown in Table 5.1, which shows the relationship between the viscosity of a formulation, the print thickness, and the degree of inhibition in cure.

Despite its very low film weight, a high-viscosity, buttery offset ink shows fairly low oxygen inhibition and is relatively easy to cure. A very low-viscosity UV inkjet, in contrast, is very difficult to cure due to rapid oxygen uptake in the film, and will offer a huge improvement in cure under nitrogen. There is an inverse logarithmic relationship between the viscosity of the formulation and the extent of oxygen diffusion into the film with the corresponding inhibition of cure. For laboratory testing purposes removed from the printing press, it is important to replicate the required viscosities in a sample to correlate the results with best practice.

Corresponding increases in photoinitiator concentrations, giving higher radical counts, are required to overcome high inhibition rates. Gravure and flexo inks will require higher concentrations than offset inks to combat oxygen uptake in the film, and inkjet demands very high levels of photoinitiators that can barely be supported by the formulation. These levels simply provide the required radical count that is necessary for full cure. For UV inkjet, the sensitized combination of ITX and Irgacure 369[10] or DETX and Irgacure 907[11] provide the very high cure speeds that are required. Similarly, with low-viscosity coatings and varnishes at 2–3 microns thickness, high levels of initiator and amines are required to achieve efficient cure.

Alternative technologies that are now being applied to very low-viscosity inkjets is cationic UV curing and hybrid curing, which do not suffer from oxygen inhibition (see Chapter 7).

High levels of amine are often used in low-viscosity, thin film UV curing to react with or scavenge the peroxy radicals by providing a source of hydrogen atoms to produce a peracid and regenerate a reactive alkylamino radical.

$$R^{\cdot} + O_2 \longrightarrow ROO^{\cdot} + CH_3N(R)_2 \longrightarrow {}^{\cdot}CH_2N(R)_2 + ROOH$$

| Active radical | Inactive peroxy radical | Tertiary amine | Active alkylamino radical | Neutral hydroperoxide |

*Figure 5.3* Oxygen scavenging by a tertiary amine.

Figure 5.3 illustrates the tertiary amine reaction with the inactive peroxy radical to produce an active alkylamino radical and a neutral hydroperoxide. The overall effect is simply one of scavenging oxygen from the system by converting it to an inactive peracid form and allowing the polymerization to take place more efficiently. Tertiary amines therefore are very efficient oxygen scavengers that will improve cure, particularly at the surface. They be used in both Type I and Type II formulations.

Thiols also react efficiently as oxygen scavengers and can be used in the same way. An alternative oxygen scavenging system that is used to good effect is a combination of benzophenone (50) with a Type I photoinitiator in the absence of an amine hydrogen donor (see Section 5.4). Benzophenone in this context is a very poor initiator, but mixtures of BP and Irgacure 184 (Irgacure 500, [80]) or BP and Darocur 1173 (Irgacure 4665 [82]) have been shown to provide enhanced surface cure. The mechanism relies on the fact that the excited triplet state of BP can decompose hydroperoxides to produce free radicals.

The hydroperoxide generated from an inactive peroxy radical/H donor source is decomposed by energy transfer from the triplet BP. The alkoxy radicals that are produced are very reactive and improve the radical count.

$$\underset{\text{triplet}}{BP^*} + \underset{\text{hydroperoxide}}{ROOH} \rightarrow \underset{\text{ground state}}{BP} + \underset{\text{alkoxy radical}}{RO^{\cdot}} + \underset{\text{hydroxy radical}}{HO^{\cdot}}$$

The triplet energy of the excited BP creates new radicals from the neutral hydroperoxide and effectively converts the oxygen that has been absorbed in the film into reactive radicals. This becomes apparent in increased surface cure in the absence of an amine.

## 5.3 Film thickness, surface and depth cure, shrinkage and adhesion

### 5.3.1 Surface and depth cure

These interrelated topics can be discussed in relation to photoinitiation at different wavelengths throughout the depth of the coating and the degree and type of cure that results. The concentration of the photoinitiator will affect many parameters such as hardness and yellowing as well as the primary cure speed.

In the polymerization process it could be said that polymerization stops when:

1. The UV source is switched off and no more radicals are generated.
2. All the monomer is consumed or gellation precludes reaction.
3. There is severe radical quenching by oxygen.
4. Chain termination occurs.

Only rarely is all the unsaturation consumed. Infrared analysis at 468 cm$^{-1}$ will show various amounts of the acrylate function remaining in the cured coating.

Rheology plays a significant role in UV curing. As the coating gels, the mobility of the reactants is reduced, making the opportunities for further reaction between the radicals and the remaining monomer less likely, and a cured coating may contain a substantial amount of residual unsaturation. Functionality also has a role, and a monofunctional acrylate is more likely to approach 100% conversion than a tetrafunctional acrylate for the same UV dose.

There are many test methods of establishing cure, and cure does not have the same meaning in all cases. Full cure is simply the point at which the desired properties of the coating are reached for a particular application.

If the molar absorption coefficient ($\varepsilon$) or the concentration of the photoinitiator is very high, then the light penetration at that wavelength through the film decreases rapidly, leading to poor depth cure.

It can be seen from Figure 5.4 that, at a fixed light dose, using 10% of initiator (by weight of the total formulation) leads to a maximum depth cure of around 8 microns thickness, whereas at 2% concentration it is possible to cure up to 25 microns depth.[12] The rate of cure of an initiator shows a rapid increase from zero up to 3 to 4% concentration then generally levels off around 6% to 7%. Above this level the rate of cure may decrease as an excess of radicals above acrylate reaction sites merely leads to radical recombination and termination reactions.

High levels of photoinitiator, despite giving higher cure speeds, will lead to reduced hardness, reduced solvent resistance, and increased yellowing. This stems from the fact that high levels of photoinitiator produce many radicals that start numerous polymer chains with low molecular weights as cure is completed. These polymers have a broader molecular weight distribution and correspondingly softer films. Low concentrations of photoinitiator, in contrast, produce a low radical count that leads to fewer, high-molecular-weight polymer species being produced,

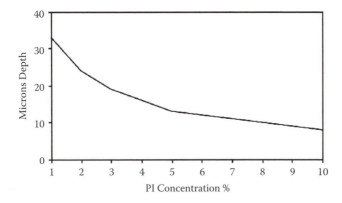

*Figure 5.4* The effect of photoinitiator concentration on depth of cure.

with a narrow molecular weight distribution and good hardness. Fairly high concentrations of photoinitiator are used to produce a high radical count and efficient curing for surface cure. Short wavelength UV has less penetrative power than long wave UV. However, the short wave output of an MPM lamp around 254 nm is sufficient to activate photoinitiators that absorb at these wavelengths, such as Darocur 1173 (1), Irgacure 184 (3) and BP (50). These initiators are used in moderately high concentrations to provide good surface cure as well as some degree of body cure (from the 313 nm output line). Short wave "germicidal" UV lamps can produce only surface cure and are sometimes used to seal the surface to prevent oxygen absorption.

When the concentration of photoinitiator is too high, all the light energy is absorbed near the surface and this will lead to rapid surface cure but poor through cure, which in turn gives rise to surface wrinkling from shrinkage and poor adhesion to the substrate.

To achieve good depth cure of thicker coatings, the MPM lamp outputs at 366 nm and 404 nm essentially provide increased light penetration.[1] Although clear composites of up to 1 cm thickness have been cured using short wave UV and 0.5% of initiators such as Darocur 1173 (1), Irgacure 184 (3) and BP/amine,[13,14] depth cure is better achieved at long wavelengths using UV A or UV-VIS. To cure at depth, low concentrations of photoinitiators such as thioxanthones and phosphine oxides are used. The low UV absorption of these photoinitiators in the long wave UV, as well as low concentrations, allow a large proportion of the light to penetrate to the substrate and influence both depth cure and adhesion. The phosphine oxides also have the advantage of photobleaching (see Section 2.5), which allows light to penetrate deeper as cure progresses and very thick coatings can be cured in this way. UV B light around 313 nm also has improved penetration, and low concentrations of photoinitiators absorbing in this area can also provide depth cure.

For very thick curing such as potting and composite formulations, 0.2% of initiators such as thioxanthones or phosphine oxides will give good results. In these applications an intermittent light source such as a multiflash xenon lamp will often give superior results. Longer wave visible light will also be useful, and 0.05% fluorenone sensitizer, absorbing at 470 nm, has been shown to give very deep cure when coupled with a phosphine oxide. For applications where thick pigmented coatings need to be cured, the doped mercury lamps with stronger outputs at these long wavelengths are beneficial. Gallium doped lamps can give increased cure of titanium dioxide whites, but the formulation needs to be balanced for these lamps.

An interesting alternative system to cure thicker coatings and reduce wrinkling employs two different lamps. A long wave fluorescent near-visible lamp emitting at 350–420 nm will provide penetrative long wave that gives body cure. This is followed by a high-pressure mercury lamp to give surface cure, providing full cure without wrinkling.

At full cure, only part of the photoinitiator will be consumed, as discussed earlier, and it is possible that over 50% of the initiator will remain unreacted. Similar effects are shown by the amine. An excess of amine unconsumed by the initiator is often used to combat oxygen inhibition, particularly in thin film curing, and at full cure there will be a large proportion of the unreacted amine left in the coating.

Surface cure can be promoted by:

1. The use of short wave UV at 200–300 nm
2. High concentration of photoinitiators that absorb in this region
3. The addition of fairly high amounts of amines, waxes, or nitrogen blanketing to reduce oxygen inhibition, particularly in thin, low-viscosity inks or coatings

Depth cure can be promoted by:

1. The use of long wave UV light at 350–420 nm
2. A low concentration of photoinitiators that absorb at these wavelengths that will allow the light to penetrate to sufficient depth
3. The use of photoinitiators that photobleach, such as phosphine oxides

In all of this, it should be remembered that a high concentration of photoinitiator will give a high radical count. This in turn will lead to the production of multiple chain polymerization reactions and fast cure, giving a low cross-link density and lower hardness values. A low photoinitiator concentration will produce a lower radical count with fewer polymeric chains, but these will have higher molecular weights and increased hardness.

## 5.3.2    Shrinkage and adhesion

Shrinkage occurs in free radical/acrylate systems as the change from double bond to single bond occurs in the polymerization.[15] Although bond length increases slightly, the change from a liquid to a solid is a more important factor and there is a loss in free volume from the much reduced mobility of the molecules. This shrinkage can be up to 20% in volume and depends very much on the degree of unsaturation in the formulation. Tri- and tetra-acrylate monomers will produce much greater shrinkage than mono- or di-acrylates. The effect of shrinkage is to produce strain in the system, which can lead to substrate warping, curl, and loss of adhesion. This is seen to a much greater extent in flexible substrates where substrate wrinkling and curl can occur. In a rigid substrate, the strain is somewhat dissipated with time. Shrinkage can be reduced by the use of high-molecular-weight oligomers or prepolymers with low unsaturation levels, but the effects depend very much on the application.

Adhesion of the coating to the substrate requires that cure is achieved at the coating–substrate interface, so depth cure must be achieved. Stress at the interface can reduce adhesion, and low shrinkage will help to promote good adhesion. In chemical terms, any photografting reactions that may occur with the substrate would be beneficial, and it has been shown that oligomers with some hydrophilic nature will promote adhesion. Chlorinated polyesters can help with adhesion as well as the use of inert resinous materials, but this is a complex topic. Several types of adhesion promoter are available commercially. These include phosphate-based products designed for metal coatings where adhesion is a particularly difficult problem.

Adhesion can be promoted by:

1. Good depth cure with a low concentration of photoinitiator at long wavelength
2. Low shrinkage from using materials with low unsaturation levels
3. The use of oligomers with some hydrophilic nature
4. The use of adhesion promoters

In addition, the surface preparation of substrates such as polyolefin stock prior to coating and cure can help with adhesion. This may take the form of a sublimation process under an arc lamp, inline corona treatment for difficult plastics, or similar techniques, but such topics are beyond the scope of this book.

## 5.4   Sensitization and synergy

### 5.4.1   Sensitization

Sensitization is the process of energy transfer between a sensitizer and an initiator. This concept is often used in pigmented UV curing where the UV absorption of the pigment is much stronger than that of the initiator and masks the amount of UV light the initiator can usefully pick up. A sensitizer is added that absorbs UV energy at wavelengths outside those of the pigment absorption, then transfers this energy to the initiator.

In Figure 5.5, the benzophenone (50), added at 2%–4%, is totally masked by the absorption at 250 nm of the blue pigment, used at perhaps 20%. All that will happen if BP is used alone is that a little surface cure will occur; there will be practically no body cure. The ITX (63) absorption at 383 nm is not masked and can pick up energy from the 366 nm and 404 nm output of the lamp and provide depth cure. Although there is little sensitization effect with ITX/BP, if the BP was replaced by Irgacure 907 in the above system, a great increase in cure speed would be seen as the sensitization effect comes into play. Using Irgacure 907 in a blue lacquer, the cure speed is increased from 5 m/min to 25 m/min by the addition of 2% ITX.[16] In a white lacquer, a similar addition has a more dramatic effect, increasing the cure speed from 2 m/min to 40 m/min.

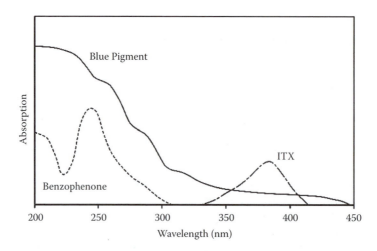

*Figure 5.5* Absorption spectra of BP, ITX, and a cyan pigment. (W. A. Green. *RadTech Eu. Conf. Proc.* 1991, 638. Reprinted with permission.)

The sensitizer S (such as ITX) absorbs UV at long wavelengths to produce an excited triplet state S*. This excited state passes on its energy to an initiator P (such as Irgacure 907) that absorbs at short wavelengths but whose absorption is masked by the pigment. The excited photoinitiator P*, which has acquired energy from the sensitizer without absorbing UV light itself, then proceeds to produce free radicals and the sensitizer returns to the ground state, ready to repeat the process in a catalytic manner.[17,18]

The initiator, which is masked by the pigment, can be used in high concentrations (say 4–5%) to provide fast cure and the sensitizer is essentially a catalyst for UV absorption and can be used in relatively small amounts (0.5–2%).

| | |
|---|---|
| S(ITX) + UV → S*(ITX*) | excited triplet sensitizer |
| S*(ITX*) + P(907) → S(ITX) + P*(907*) | triplet – triplet energy transfer to initiator |
| P*(907*) → R· | initiator produces radicals |

Sensitizers can also be used in applications other than inks since the process often leads to very efficient photoinitiation..

For energy transfer to take place, the triplet energy of the sensitizer must be greater than that of the initiator. Materials that are photoinitiators themselves can be used catalytically as sensitizers provided there is a positive triplet energy balance in favor of the sensitizer. Sensitization effects are generally stronger in a more polar medium.

Thioxanthones such as ITX (63) absorb in the long wave UV at 370–410 nm and have a high triplet energy of 61–63 kcal.$M^{-1}$. They are most often used as sensitizers in inks where the pigment absorption can be avoided.

Hydroxyacetophenones such as Darocur 1173 (1) and Irgacure 184 (3), and ketals such as Irgacure 651 (14), have higher triplet energies of 63–71 kcalM$^{-1}$ and cannot be sensitized by thioxanthones through energy transfer.

Alkylaminoacetophenones, such as Irgacure 907 (10), and Irgacure 369 (12) and 379 (14), have lower triplet energies of 60–61 kcal.$M^{-1}$ and are readily sensitized. These initiators are often combined with thioxanthones for high-speed UV inks.

Benzophenone (50) has a high triplet energy of 69 kcal.$M^{-1}$ and is difficult to sensitize but is an excellent sensitizer itself, the only drawback being that BP absorbs in the short wave UV and will have little effect in inks due to the strong pigment absorption. Substituted BPs such as Speedcure PBZ (58) and Speedcure BMS (60) and phosphine oxides[20] such as Lucirin TPO (22), TPO-L (23) and Irgacure 819 (24) have lower triplet energies and can be sensitized by ITX to some degree. Some optical brighteners have been

*Table 5.2* Cure Speed Sensitization by ITX

| Irgacure 907 % | Irgacure 369 % | ITX % | Cure speed m/min |
|---|---|---|---|
| 2 | | | 20 |
| 2 | | 0.5 | 50 |
| | 4 | | 33 |
| | 4 | 1 | 130 |

shown to improve the cure when phosphine oxides TPO and 819, are used and can be regarded as sensitizers.[21] In all of these combinations, the rate at which energy transfer takes place will determine the effectiveness of the sensitization process. For example, small amounts of ITX will greatly increase the cure speed of Irgacures 907 and 369,[17–19] as shown in Table 5.2.

Other long wave triplet sensitizers such as anthracene, perylene, fluorenone, etc., have been used academically.

These combinations of thioxanthones and alkylaminoacetophenones are used in high-speed litho and flexo inks. A combination of, say, 3% Irgacure 369, 2% Speedcure ITX and 3% Speedcure EDB gives one of the fastest systems available. The ITX provides the long wave absorption, the 369 provides very fast cure by sensitization, and the amine EDB gives added speed by providing surface protection through oxygen scavenging.

Short wave sensitizers with high triplet energies, such as acetone and acetophenone, can be used to increase the reactivity of clear coatings but would be masked themselves by pigments and have little effect in UV inks. A polymeric aminoacetophenone, Omnipol SZ, has been developed that absorbs at 320 nm. Omnipol SZ will sensitize benzophenone and its derivatives, such as 4-phenylbenzophenone (PBZ), and provide improved cure in the darker colored inks as well as low migration properties.

In the mixtures of BP and trimethylbenzophenone (TMBP) such as Esacure TZM (80) and Esacure TZT (99), BP acts as a sensitizer from its excited triplet state. This energy is transferred to the TMBP, which becomes an excited triplet and reacts with the hydrogen donor. Both BP and TMBP can react as initiators on their own, but the combination is more efficient than the individual photoinitiators and provides a little extra cure speed. Interestingly, the absorption wavelengths are very similar at 248 nm and 245 nm so this is not a system that could be usefully employed in inks but is a more efficient system for clear coatings.

## 5.4.2  Synergy

Synergy generally refers to a mixture of two or more products that produce an advantage in photoactivity compared with the individual materials used alone, and synergistic mixtures are common in UV curing. Sometimes a sensitization effect occurs to produce a more efficient mixture. Alternatively, a cage-type complex that improves efficiency may be formed. The latter can occur if a mixture of photoinitiators is ground together prior to adding to a formulation and a small increase in photoactivity may be observed.

A 50/50 mixture of BP and 184 is sold commercially as Irgacure 500 (81). This system provides some degree of oxygen scavenging and produces better surface cure than either of the products individually (see Section 5.2).

## 5.5    The effect of pigments on the UV curing process

### 5.5.1    Pigment absorption and the pigment window: Choice of photoinitiator

Pigments absorb discrete parts of the visible spectrum and the remaining light is then reflected, conveying color. Pigments also absorb light in the UV spectrum, particularly at the shorter wavelengths. This has an adverse effect on the curing process in that UV absorption by the pigment, sometimes loaded at 20% of the formulation weight, will leave little UV energy available for the photoinitiator, which may be loaded at only 2%. In the worst case, UV light will not be able to penetrate to any depth due to pigment absorption and curing will take place only at the surface.[22-25]

The transmission spectrum of a pigment will show a low UV absorption and high transmission at one part of the spectrum, say 300–350 nm. This is termed the "pigment window," where light is more readily transmitted and where the absorption of a photoinitiator will be least affected by pigment screening and is likely to be most efficient. To maximize the UV reactivity of a formulation, photoinitiators that have a high absorption in the "transmission window" of each pigment used should be chosen.

The wavelengths of the pigment "window" will vary with the color and pigment type, but in general are

Magenta: 300–400 nm
Yellow: 290–370 nm
Cyan: 370–400 nm
Green: low transmission throughout UV
Black: low transmission throughout UV
Titanium dioxide white: >380 nm

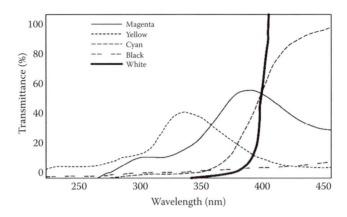

*Figure 5.6* Transmission spectra for pigments.

It can be seen from the transmission curves shown in Figure 5.6 that very little short wave UV light below 280 nm is available from any of these pigments. The binders used, urethane acrylates and epoxy acrylates, also absorb to some extent in the short wave UV below 240 nm. Short wave initiators such as Darocur 1173 (1) or BP (50) are unsuitable for inks when used alone and will only lead to surface cure with little depth cure or adhesion. The transmission curve for pigment yellow shows, for example, 30–40% transmission at 310–360 nm. Photoinitiators such as Irgacure 369 ($\lambda_{max}$ 320 nm) and Speedcure BMS ($\lambda_{max}$ 312 nm) absorb UV in this region and can be used effectively to cure yellows.

Certain combinations of pigments are difficult to cure because one pigment may cancel out the transmission window of the other color. If titanium dioxide white is added to the yellow, it will nullify the yellow pigment window since $TiO_2$ transmits very little below 380 nm, and above this wavelength pigment yellow has a strong absorption itself. This combination will therefore be very difficult to cure. Similarly, mixtures of pigments such as cyan and yellow to form a green will be difficult to cure since the cyan blocks the yellow pigment window and the mixture allows little light to penetrate the film.

Some pigments may be incompatible with certain photoinitiators. In particular, basic pigments can affect the reactivity of phosphine oxides in relation to formulation stability in that the phosphine oxides are particularly susceptible to hydrolysis in the presence of nucleophilic materials.

The four colors in a set of process inks—magenta, yellow, cyan and black—will give different cure speeds if all are formulated with the same initiator package and concentration. The degree of difficulty of cure increases in this same order of color. Initiators need to be formulated in different concentrations and different mixtures to allow each color to cure at a similar speed. This is essential to provide full cure at the same press speed over the wide range of colors that are formed from the process colors.

An alternative approach is to formulate a robust photoinitiator package that will absorb throughout the UV spectrum, and use this in different concentrations for the different colors to provide similar rates of cure.

When considering the absorption of a photoinitiator it is useful to take note of some secondary absorption peaks rather than just the main absorption peak. For example, BP (50) has a small but very useful absorption at 345 nm tailing off to 375 nm. When the main absorption peak at 245 nm is masked by a pigment, this secondary peak can be activated by the strong 313 nm or 366 nm lamp output or even by the minor UV output at 334 nm, allowing some degree of depth cure.

Tailing of the initiator absorption toward the long wave can also be useful. Irgacure 907 (10) has a maximum absorption at 304 nm and is most closely associated with the 313 nm output line but the tail continues absorbing up to 380 nm and can utilize the small 334 nm output and the strong 366 nm line. Similarly, Irgacure 369 (12), absorbing at 324 nm, tails off into the visible range up to 420 nm and can pick up the important 404 nm output as well as the 366 nm line.

Mixtures of photoinitiators will often cure much better than a larger amount of a single photoinitiator. For example, in a cyan ink, 2% of Irgacure 369 plus 2% of Irgacure 184 (or 2% Irgacure 651) will cure twice as fast as 4% of Irgacure 369 alone. This is mainly a surface curing effect of the 184 or 651.

## 5.5.2   Replacing ITX

Recent migration problems with ITX have led many formulators to produce inks for particular applications without using ITX. This is not an easy task since most UV inks have long relied on ITX as a long-wave photoinitiator for the darker colors.

First, we have to look at what ITX provides, other than being just a radical source:

- 360–420 nm long wave UV absorption for inks
- Low molar absorption, E(1%, 1 cm) 168, for depth cure
- High triplet energy (61.4 kcal/mol) for sensitization and high speed cure
- Low molecular weight of 254.3 leads to migration
- Cost-effective product

Replacements that fit the long wave UV absorption of ITX can be found only in the phosphine oxides TPO and Irgacure 819. Phosphine oxides are, however, very sensitive to oxygen inhibition and are inefficient in thin films such as litho and flexo inks, which is one of the main markets for ITX. As a sensitizer, using small amounts of ITX with any alkylamino-acetophenone can produce very high cure speeds from the transfer of triplet energy to Irgacures 907 and 369 (Chapter 5.4.1). The phosphine oxides have lower triplet energies (60 and 55 kcal/mol) and are unable to sensitize AAAPs or to provide very fast cure. Polymeric thioxanthones retain the long wave UV absorption and high triplet energies and are technically good replacements for ITX, but their higher costs and reformulation make them insufficiently cost effective in many cases.

Irgacure 369 (300–350 nm) is a very efficient initiator but does not absorb quite far enough into the near visible to give the same performance as ITX in dark inks. Many combinations of photoinitiators have been promoted for inks. Irgacure 369 and Irgacure 651 in ratios of about 1:2 (Irgacure 1300 blend) can provide a fast-cure package but Irgacure 651 with its odor cannot be used in packaging inks, and the combination yellows considerably. Ethyl anthraquinone (310–360 nm) absorbs similarly to 369 but has a lower triplet energy of around 60 kcal/mol and is essentially a less efficient photoinitiator than ITX or 369.

In brief, ITX is cost effective and possesses unique properties as a photoinitiator and sensitizer and there would appear to be no direct replacement.

In practice, there are many factors other than reactivity that will influence the choice of a photoinitiator such as type of mechanism and hydrogen donor used, solubility and ease of formulation, odor and taint from unwanted by-products, yellowing, hazard labeling, handling restrictions, cost effectiveness, etc., most of which are influenced by the application.

## 5.5.3   Titanium dioxide whites and carbon blacks

Titanium dioxide in its rutile or anatase form absorbs throughout the UV spectrum up to 390 nm and offers no UV pigment window below this wavelength. To cure formulations that contain titanium dioxide, initiators must be chosen that absorb over 390 nm and into the near visible. The thioxanthones, plus a low yellowing amine, can be used, such as ITX (63) and Speedcure DMB (114), but thioxanthones are inherently yellow since they absorb in the near visible and this leads to some yellowness in the cured coatings if the concentration of thioxanthone is too high.

The phosphine oxides, Lucirin TPO, TPO-L and Irgacure 819 (22–24), absorb up to 420 nm, are very efficient, and have become widely used for this application. The photobleaching qualities of these initiators ensure that the initial yellowness of the formulation is lost with exposure time and leads to excellent whites. The phosphine oxides are very sensitive to oxygen inhibition and need to be protected by the addition of surface curing photoinitiators such as Darocur 1173 (1) or Irgacure 184 (3). Commercial blends of TPO and 1173, and of 819 and 1173 in ratios of about 1:4, are often used for white inks and unsaturated polyester coatings and now dominate the "whites" market.

Carbon black absorbs UV and visible light at all wavelengths and allows little UV to penetrate into the coating and be absorbed by the photoinitiator. The transmission of UV is no more than 10% in the near visible range at 350–450 nm. Hence, the curing of carbon black inks is something of a problem area. It is essential to use only sufficient carbon black to provide the color density that is required and then use a cocktail of photoinitiators, such as ITX (63), Irgacure 651 (14), and Irgacure 907 (10), or similar, that will absorb UV over most wavelengths to achieve cure.

Formulating inks and coatings is a complex art beyond the scope of this book and requires many factors to be taken into account other than just the pigment absorption to produce an efficient formulation. These factors may include:

- The type of resins and monomers that are used including their functionality.
- Fillers that may be used.
- Pigment type. Some pigments may not be suitable for the application. Different types of pigments will be required for applications such as food packaging, outdoor posters, wallpapers, etc. Color fastness may be an important factor.
- Particle size. Light scattering from the pigment particles is a significant effect in improving cure and particle size, and shape can be important. Wetting properties and rheology may be affected. Metallic effects or "plate" forms of pigment are often difficult to cure.
- Substrate type and color. UV reflection from the substrate improves cure and adhesion. Colored substrates, black in particular, will reflect little UV and adhesion will be more difficult to achieve.
- Screening effects of other additives such as the UV absorption of the aminobenzoate hydrogen donors around 310 nm. Aromatic oligomers absorb strongly below 300 nm and can also mask the photoinitiator.

- Matting effects. Although silica particles used for matting effects are UV transparent, they will affect rheology and need to be taken into account.
- Yellowing during and after cure.
- The solubility of the photoinitiator, affecting the stability of the formulation.
- The film thickness and the required cure speed, screen, litho, flexo inks, etc.
- Specific lamp emissions and the associated power outputs.
- Relative costs of the initiators.
- Toxicity and labeling for certain applications.

## References and further reading

1. de Ruiter, B. (TNO Plastics and Rubber Res. Inst.) Toward optimised combinations of UV lamps and photoinitiators. *RadTech Eu. Conf. Proc.* 1991. Paper 47, 597–603.
2. Chang, C-H., A. Mar, H. Evers, D. Wostratzky. (Ciba) The effect of light sources and photoinitiators on through cure in pigmented systems. *RadTech NA. Conf. Proc.* 1992. 16–27.
3. Brandl, B. (IST Metz) UV-LEDs. Survey of a new engineering technology. *RadTech Eu. Conf. Proc.* 2007.
4. Hanrahan, M. J. (EM Industries) 1990. Oxygen inhibition: Causes and solutions. *RadTech Report*, March/April, 14–19.
5. Hanrahan, M. J. (EM Industries) 1992. Overcoming the effects of oxygen on cured and uncured UV formulations. *J. Rad. Curing.* Spring, 40–47.
6. Decker, C. (CNRS, Mulhouse) UV Curing chemistry: Past, Present and Future. *RadTech Eu. Conf. Proc.* 1987, 1–7 to 1–25.
7. Wallis, R. (Wallis Surface Science) The economics and performance of UV curing with inert gas in the production of furniture foils. *Proc. PRA Economy and Performance Conference.* Manchester (UK), 2004, Paper 4.
8. Rames-Langlade, G. (Air Liquide SA) Gas sealed UV Dryer for optimising UV applications. *RadTech Eu. Conf. Proc.* 2005. 59–65.
9. Biehler, M., E. Beck, K. Menzel, S. Titusson, A. Daiss. (BASF) K. Soljauer, K. Fageholm. (Tikkurila Coatings) UV becomes 3-dimensional. *RadTech Eu. Conf. Proc.* 2005. 59–65. ($CO_2$)
10. Fuchs, A., M. Richert, S. Biry, S. Villeneuve, T. Bolle. *RadTech Report*, 2004, Sept./Oct. 41–45. UV Ink-jet. A. Fuchs, M. Richert, S. Villeneuve. PCT Int. Patent Appln. WO 92, 287. 2004.
11. Overend, A. S., J. Cordwell, M. Parry. PCT Int. Patent Appln. WO 26,978, 2004. A. Hayashi. EP 1,426,421. 2004. UV ink-jet.
12. Hanrahan, M. J. (EM Industries) The effect of photoinitiator concentration on the properties of UV formulations. *RadTech NA. Conf. Proc.* 1990. 249–256.
13. Van Landuyt, D. C. (Celanese Plastics) Practical aspects of thick film UV curing. *J. Rad. Curing.* July 1984, 4–8. (pre-TPO)
14. Aldridge, A. D., P. D. Francis, J. Hutchinson. (Elec. Council Res. Centre) 1984. UV curing of $TiO_2$ pigmented coatings. *J. Oil Col. Chem.Assoc.* 2, 33–39.

15. Holman, R. 1998. Shrinkage. *PRA. RADnews* Autumn (26), 28.
16. Dietliker, K., M. W. Rembold, G. Rist, W. Rutsch, F. Sitek. (Ciba) Sensitization of photoinitiators by triplet sensitizers. *RadTech Eu. Conf. Proc.* 1987. 3–37 to 3–56.
17. Fouassier, J. P., D. Ruhlmann. (ENSC Mulhouse) How to understand synergistic phenomena in the photopolymerization of pigmented systems. *RadTech Eu. Conf. Proc.* 1991. Paper 39, 500–509.
18. Rist, G. A., Bover, K. Dietliker, V. Desobry. (Ciba) 1992. Sensitization of aminoketone photoinitiators. *Macromolecules* 25, 4182–4193.
19. Fouassier, J. P., D. Ruhlmann, B. Graff, F. Wieder. (ENSC Mulhouse) 1995. New insights in photosensitizer–photoinitiator interaction. *Prog. Org. Coat.* 25, 169–202.
20. Williams, R. M., I. V. Khudyakov, M. B. Purvis, B. J. Overton, N. J. Turro. (Columbia Univ. NY) 2000. Direct and sensitized photolysis of phosphine oxide photoinitiators. *J. Phys. Chem. B.* 104, 10437–10443.
21. Wilczac, W. A. Fluorescent whitening agents in UV curing. *RadTech NA. Conf. Proc.* 2000, 570–576.
22. Grierson, W. (Ciba) 1998. Selection of pigments for use in UV curing inks. *PRA RADnews.* Winter (27), 1–5.
23. Herlihy, S. L., G. C. Battersby. (Coates Lorilleux) UV inner filter effects and photoinitiation efficiency. *RadTech NA. Conf. Proc.* 1994. 156–168.
24. Ilg, S., M. Kunz, T. Bolle, R. Schulz, V. Sitzmann. (Ciba) Photoinitiator selection for UV flexo inks. *RadTech NA. Conf. Proc.* 2000. 402–413.
25. Menzel, K. (BASF) Latest investigations in formulation and processing of pigmented UV coatings. *Proc. PRA Economy and Performance Conference,* Manchester (UK), 2004. Paper 12.

*chapter six*

# Photoproducts and the yellowing of coatings

There are many reasons that a cured coating may display some yellowing. There may be intrinsic color shown from some of the additives, such as thioxanthone photoinitiators, and the formation of yellow photoproducts may occur from the decomposition or rearrangement of the radical species. Tertiary amines tend to bring long-term yellowing[1] due to atmospheric oxidative processes, particularly with outdoor applications.

## 6.1   Intrinsic color, photoyellowing, and oxidation products

Photoinitiators that absorb light above 400 nm and do not photobleach, such as thioxanthones, are yellow. Not all of the photoinitiator is consumed during cure and the remaining thioxanthone may impart some color. Although thioxanthones are widely used in ink formulations, their yellow color is generally masked by that of the pigments, unless delicate pastel shades are used. Up to 0.5% of thioxanthone is occasionally used to give some cure boost to a clear varnish, but amounts above this level will impart some color.

Over time, tertiary amines undergo photo-oxidative processes that produce yellow products such as enamines and amides. The degree of yellowing varies, depending on factors such as the amine structure and the exposure to regimes such as strong sunlight and weathering. The aminobenzoate esters tend to yellow more than simple tertiary amines or oligomeric amines.

The photoyellowing that occurs[1,2] will be proportional to the UV dose that is used during cure and the film thickness. A small amount of yellowing can also come from the oligomers that are used, and aliphatic urethane acrylates tend to give less yellowing than aromatic prepolymers or polyethers. The use of stabilizers can reduce the yellowing effect.

## 6.2   The formation of photoproducts

The free radicals that are formed during the UV curing process are required to start the polymerization of the acrylate monomers and oligomers. A large excess of radicals is generally produced to give a good cure speed and counter the effects of oxygen inhibition. The radicals that react become part of the growing polymer chain. Some of these excess radicals may undergo further scission, rearrangement, and combination reactions, mostly thermally induced, to produce a wide variety of new products. Some of these photoproducts, although present in very small amounts, may be strongly colored or odorous. These numerous photoproducts are mostly small molecules that will also be prone to migration under appropriate conditions.

## 6.2.1   Photoproducts from Type I photoinitiators

Type I photoinitiators tend to give a wide variety of low-molecular-weight photoproducts since the scission process produces two radicals that can recombine with each other in different ways as well as interacting with other radical species. Further scission of the main radicals may also occur, producing other radicals and photoproducts. GCMS analysis of a UV-cured coating can indicate a dozen or so photoproducts from a single photoinitiator.

The typical Type I photoinitiator, 2,2-dimethyl-2-hydroxyacetophenone Darocur 1173 (1), produces a benzoyl and an alkyl radical that are both reactive and will initiate the polymerization. Figure 6.1 shows some of the subsequent reactions of these radicals and the photoproducts that are produced.

The benzoyl radical, in the presence of a hydrogen donor, will produce benzaldehyde, which is a common photoproduct of most Type I photoinitiators and may lead to odor. Oxidation of this gives benzoic acid. Two benzoyl radicals in recombination give benzil, which is yellow. Alternative recombination leads to a semi-quinoid structure that is also yellow and somewhat unstable. Photoproducts coming from the quinoid structure can lead to extended chromophores that again may be yellow. In the case of 1173, these photoproducts occur only in very small traces and lead to very little yellowing.

The alkyl radicals associated with Type I photoinitiators are varied and numerous. Many of the alkyl radicals undergo thermal rearrangements to produce other radical species that may also be reactive. These may include some odorous and yellow by-products. In the case of 1173, the alkyl radical that is first produced disproportionates to give isopropanol and acetone, the latter being the main photoproduct. A small amount of glycol may also be formed.

All these products from the 1173 alkyl radical are colorless and volatile and are generally driven off under the heat of the curing process. This factor makes the HAPs most suitable for very low yellowing coatings.

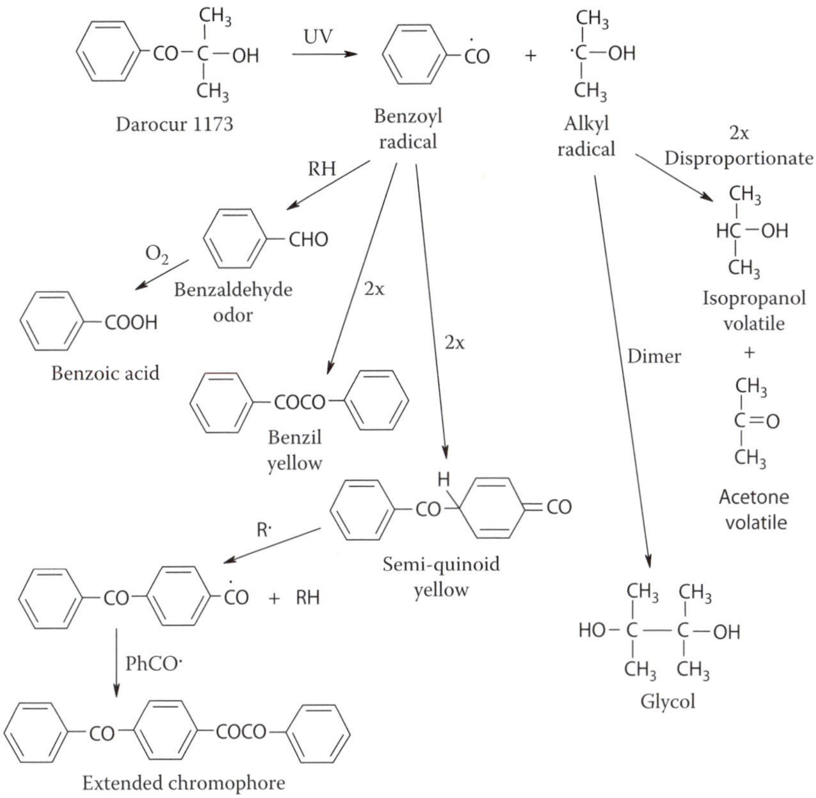

*Figure 6.1* Photoproducts from Darocur 1173.

The formation of photoproducts from benzildimethyl ketal, Irgacure 651 (14), is well documented[3] and Figure 6.2 further illustrates these points.

The benzoyl radical follows the same pattern, producing benzaldehyde, benzoic acid, benzil, quinone, and extended chromophores. In addition, recombination of the benzoyl and benzyl radicals gives another quinoid structure, which is unstable and probably is responsible for the initial yellowing that occurs after cure. This fades and leads to less-yellow products.

The benzyl radical undergoes rearrangement to produce methyl benzoate, which is the main photoproduct and produces odor, the main reason that Irgacure 651 is not used in food packaging and applications where taint is a problem. An offshoot of the rearrangement of the benzyl radical is a methyl radical, which is very reactive and will be found to give much of the photoactivity from Irgacure 651. The methyl radical may also combine with the benzoyl radical to produce acetophenone.

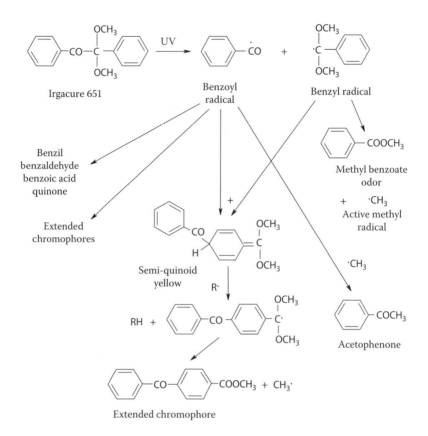

***Figure 6.2*** Photoproducts of 2,2-dimethoxy-2-phenylacetophenone, Irgacure 651.

Irgacure 907 (10), 2-methyl-4'-(methylthio)-2-morpholinopropiophenone, shows some degree of β-scission (Figure 6.3), leading to different photoproducts, as well as the standard (substituted) photoproducts from the main α-scission process.

Alpha-cleavage produces a substituted benzoyl radical and an alkylamino radical, both of which are very reactive and will initiate polymerization. Beta-cleavage produces a phenacyl radical and a morpholino radical (Figure 6.3).

In the major Type I scission process of Irgacure 907, illustrated in Figure 6.4, the substituted benzoyl radical leads to typical photoproducts such as methylthiobenzaldehyde and benzil. The alkylamino radical can also interact with the benzoyl radical.

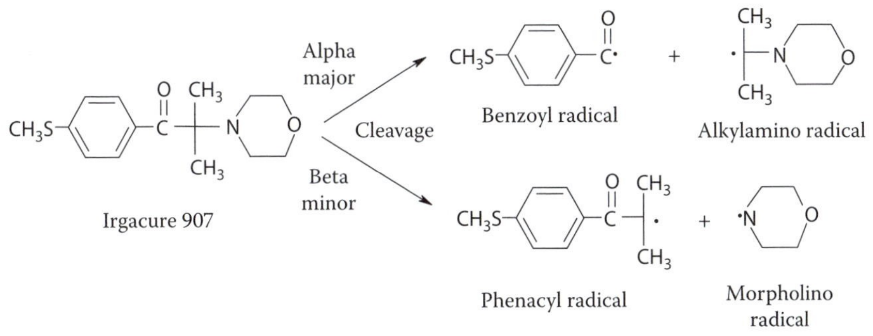

*Figure 6.3* Alpha and beta cleavage of Irgacure 907.

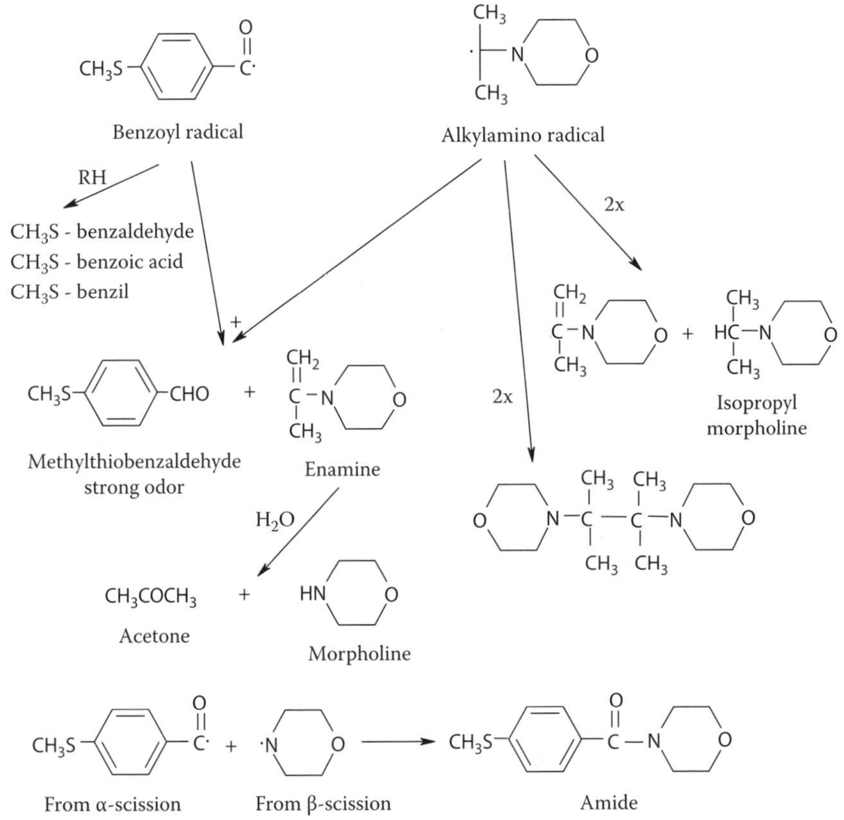

*Figure 6.4* Photoproducts of Irgacure 907.

From α-cleavage, the substituted benzoyl radical (by hydrogen abstraction) produces methylthiobenzaldehyde, which is a volatile liquid with a strong odor. This is the main photoproduct and is immediately noticeable by its strong sulphury odor after cure. Reaction of the two main radicals, benzoyl and alkylamino, can also produce this product together with an enamine. The latter, in the presence of water, will produce acetone and morpholine. The alkylamino radical can also disproportionate to give an enamine and isopropylmorpholine, as well as forming a dimer. Reaction of the benzoyl radical from α-scission with the morpholino radical from β-scission can lead to an amide being formed.

Irgacure 369 (12), 2-benzyl-2-(dimethylamino)-4'-morpholinobutyrophenone, will release 4-morpholinobenzaldehyde in a scheme similar to the previous, but this material is much less volatile or odorous than the aldehyde from Irgacure 907. The corresponding alkylamino radical from 369 can lead to yellow photoproducts. Minor β-cleavage of the C-N bond also leads to the dimethylamino radical. This may take up a hydrogen atom giving dimethylamine or may recombine with the benzoyl radical, producing p-(morpholino)-N,N-dimethylbenzamide.[4] Irgacure 379 (14) reacts very similarly to 369.

The photoproducts of 1-hydroxycyclohexyl phenyl ketone (Irgacure 184), (3), photolyzed in methanol, have been studied by GCMS. Cyclohexanone is perhaps the most abundant photoproduct. Other photoproducts of 184 include cyclohexanol, 2-hydroxycyclohexanone, 2-hydroxy-1-phenylethanone, 1,1-dimethoxycyclohexane, methyl benzoate, 2-diphenylethanone, benzaldehyde, benzoic acid, benzil, and O-benzoylbenzoin. Most of these products are found in only trace amounts.

The phosphine oxides Lucirin TPO (22), TPO-L (23) and Irgacure 819 (24) produce a 2,4,6-trimethylbenzoyl radical and a phosphinyl radical that is equally, if not more, reactive. Traces of 2,4,6-trimethylbenzaldehyde from hydrogen abstraction by the benzoyl radical have been detected. Transient photoproducts from the phosphine oxides initially give a reddish color but this bleaches with exposure. There are few yellow or odorous photoproducts from the phosphine oxides.

It would be difficult to record the photoproducts of all the commercial photoinitiators but the schemes outlined illustrate the nature of the breakdown of a Type I photoinitiator through photoscission, thermal abstraction, rearrangement, and combination reactions during the UV-curing process. Most photoinitiators will produce only one or two main photoproducts and the majority of the other photoproducts are found in only trace amounts.

Following the (substituted) benzoyl radical and the (appropriate) benzyl or alkyl radical in such schemes as those shown will indicate the nature of the possible photoproducts that may be produced from most of the Type I photoinitiators.

## 6.2.2 Photoproducts from Type II photoinitiators

Benzophenone (50) plus a tertiary amine illustrates a typical Type II reaction scheme in Figure 6.5, producing photoproducts mainly by radical interactions. The initial reaction under UV excitation produces a ketyl radical from the benzophenone, which is inactive, and an alkylamino radical from the tertiary amine, which is very reactive. Unlike the Type I system, there are no scission products and fewer photoproducts in general.

The ketyl radical (in Figure 6.5) forms a pinacol dimer, the main photoproduct, and there are some termination reactions that limit the growing polymer chain. Some disproportionation of the ketyl radical also occurs.

The alkylamino radical is the prime initiator species which, like all radicals, can be quenched by oxygen, forming an inactive peroxy radical. Interaction of the peroxy radical with excess tertiary amine produces another active alkylamino radical plus a hydroperoxide. The latter can lead to long-term yellowing but, being aliphatic, this is not too significant.

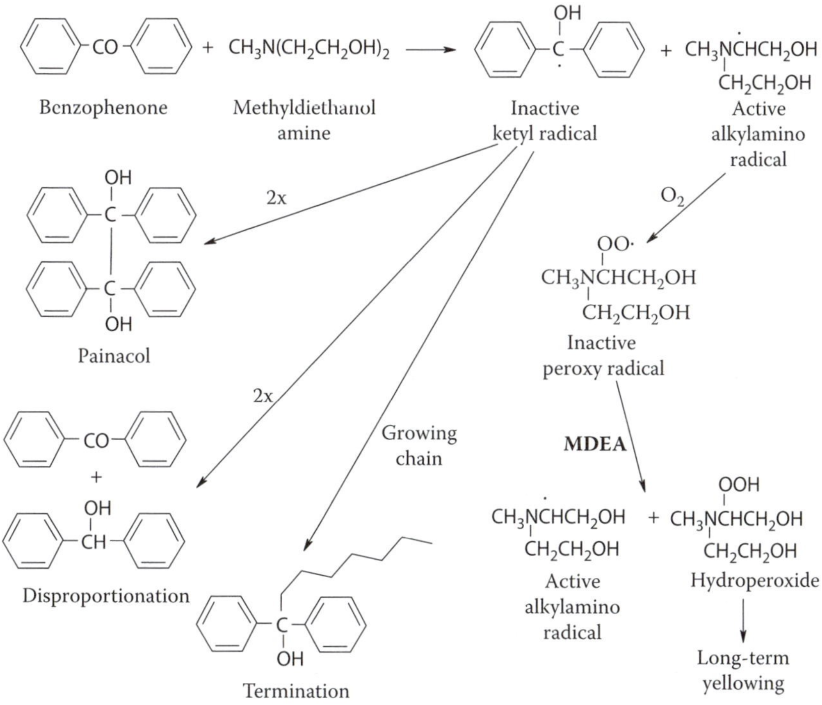

**Figure 6.5** Photoproducts from benzophenone and methyldiethanolamine.

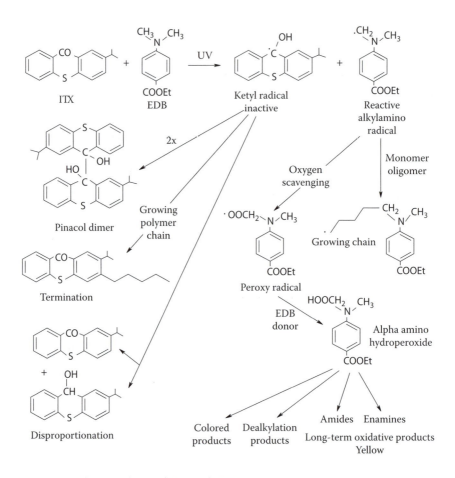

***Figure 6.6*** Photoproducts of ITX and EDB.

A Type II scheme is outlined in Figure 6.6 for the photoproducts of isopropylthioxanthone, ITX (63), and the aminobenzoate Speedcure EDB (115) follows similar lines.

The ITX (63-72) ketyl radical, like most Type II ketyl radicals, has a poor reactivity toward acrylates and will mainly provide a source of termination, reacting with the growing polymer chain. Dimerization to produce a pinacol is limited due to steric effects, and disproportionation takes place to some extent, producing isopropylthioxanthone. Many of these species derived from the ketyl radical are longer-chain thioxanthones and will be yellow photoproducts.

The tertiary amine, Speedcure EDB (115), produces a very reactive alkylamino radical that will initiate polymerization, becoming part of the growing chain. EDB can also react with any inactive peroxy radicals that have been formed by oxygen quenching, regenerating an active radical and a hydroperoxide. The long-term oxidation of these hydroperoxides and similar species will lead to some yellowing of the coating with time. Aromatic tertiary amines such as aminobenzoates tend to lead to more yellow photoproducts than aliphatic amines.

Other types of tertiary amines will react similarly to EDB, producing radicals that initiate the growing chain, scavenge oxygen, and produce long-term photoproducts that may be yellow.

p-Tolylthiobenzophenone, Speedcure BMS (60), produces no significant color. Phenylbenzophenone, Speedcure PBZ (58), can produce some yellowing in its photoproducts through the conjugated aryl species.

Methyl benzoylformate, MBF (73), cured in air in a urethane acrylate, gives very little yellowing. Oxygen suppresses the biradical from further scission but, under nitrogen, some yellowing occurs due to the formation of phenylglyoxylic acid and other products.

## References

1. Arsu, N., R. S. Davidson, R. Holman. 1995. Factors affecting the photoyellowing which occurs during the photoinitiated polymerization of acrylates. *J. Photochem. Photobiol. A: Chem.* 87, 169–175.
2. Studer, K., R. Koniger. (BASF) Initial photoyellowing of photocrosslinked coatings. *Eur. Coat. J.* 1–2, 2001, 26–36, 57, 58.
3. Kirschmayr, R., G. Berner, R. Huesler, G. Rist. 1982. *Farbe Lack,* 88, 910.
4. Desobry, V., K. Dietliker, L. Misev, M. Rembold, G. Rist, W. Rutsch. Novel photoinitiator for modern technology. In *Radiation Curing of Polymeric Materials;* C. E. Hoyle, J. F. Kinstle. Eds. ACS. 1990, 92.

# chapter seven

# Cationic chemistry

Cationic curing is a relatively small but growing area of interest in the UV curing industry, mainly due to the inherently more expensive materials that are used compared with free radical systems, but it brings a new set of parameters and a different chemistry to widen the scope and application of UV curing. Cationic curing involves the use of UV energy to generate a protonic acid from the cationic photoinitiator. Unlike free radicals, which promote a chain polymerization of acrylate-type monomers, the protonic acid will initiate a ring opening polymerization of epoxy resins, forming polyethers. These resins bring properties to UV-cured coatings that differ from those of acrylates, and cationic curing has found several niche markets such as metal decorating, adhesives, and inks where the advantages of excellent adhesion, insensitivity to oxygen, and good chemical resistance of the polymer are required. Cationic curing, compared with free-radical curing, provides coatings that

- Are insensitive to oxygen during cure
- Give very low shrinkage
- Give excellent adhesion and chemical resistance
- Are subject to thermal post-cure
- Produce a more uniform and stable polymer
- Are of an acidic nature

The process is easily inhibited by trace amounts of basic materials such as amines, urethanes, fillers, and some basic pigments such as magenta 57:1.

Early photoinitiators of this nature involved aryldiazonium salts that produced Lewis acids under UV. These photoinitiators were inherently unstable thermally sensitive materials and produced nitrogen in the curing process, all features that were detrimental to their commercial use. In today's UV industry, cationic photoinitiators are mostly onium salt structures such as triphenylsulphonium and diphenyliodonium salts coupled with a non-nucleophilic anion. This type of anion ensures that the Brønsted acid formed is stable and that the anion has little tendency to take part in any termination reactions. Ferrocenium salts offer an alternative source of Lewis acid but are generally less efficient than the onium salts.

## 7.1  The light absorption process and the generation of acid

The use of UV energy to produce a protonic acid from the onium salt follows a similar pattern to the light absorption processes that produce free radicals. The onium salt absorbs UV light producing an excited singlet state. Cleavage of the carbon-sulphur or carbon-iodine bond then produces an aryl cation. In the presence of a hydrogen donor (from polyethers or polyols commonly used in the formulation), the aryl cation produces a proton which, coupled with the anion, provides the Brønsted acid.[1-3]

### 7.1.1  Triphenylsulphonium salts

In the case of a triphenylsulphonium salt, heterolytic cleavage is the dominant pathway and these reactions are illustrated in Figure 7.1:

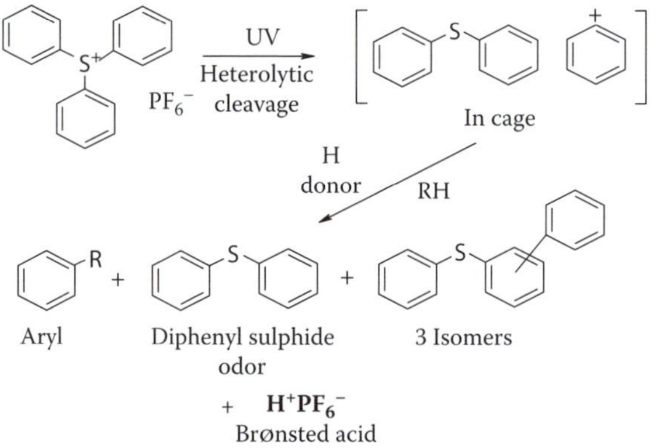

*Figure 7.1* The formation of a Brønsted acid from heterolytic cleavage of a sulphonium salt.

An aryl cation is first formed in a cage. The release of a proton, which forms the Brønsted acid via the hydrogen donor, produces a substituted aryl, mainly bound to the oligomer. Diphenyl sulphide and the three isomers of phenylthiobiphenyl are the main photoproducts, all bringing odor and the characteristic sulphury smell, which is very noticeable immediately after cure.

Other minor photo and thermal reactions lead, via homolytic cleavage, to different by-products, some of which may be significant to the curing process and the application (Figure 7.2). For example, an unsubstituted

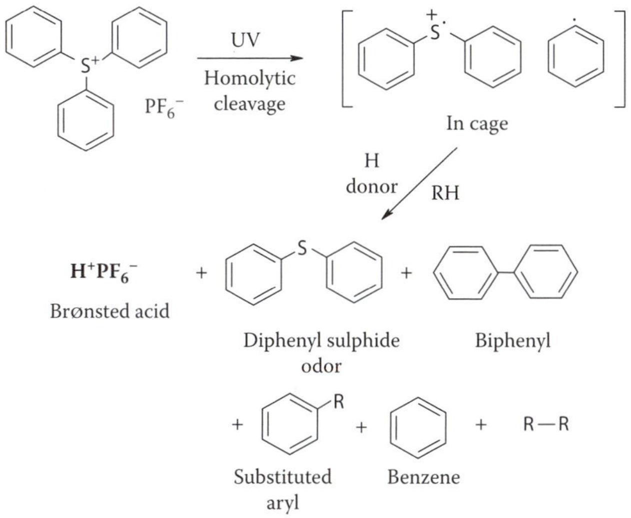

*Figure 7.2* Minor homolytic cleavage of a sulphonium salt.

aryl group in the cationic salt will produce a phenyl radical. This may lead to the formation of benzene from hydrogen abstraction and biphenyl from radical–radical interaction. In UV curing, a high density of reactive species is produced and radical interactions, rather than abstraction, are more likely. While benzene may be formed in trace amounts, it is a significant toxicological hazard for some applications. However, benzene should be considered in the context of a variety of exposure scenarios both natural and synthetic, being present, for example, in commonly handled substances such as gasoline.[4] Biphenyl, a registered food antioxidant, is considered harmless. Subsequent thermal treatment, sterilization, etc., of products such as metal cans will remove these photoproducts and any other odorous by-products. Despite the release, in some cases, of these odorous photoproducts, cationic sulphonium salts have been assessed for safety and migration under FDA tests and found suitable for printing inks for polyolefin stock and food packaging applications.[5]

Other by-products that may be produced from radical coupling and hydrogen abstraction include higher substituted aryl sulphides and traces of some colored products.

## 7.1.2 *Diphenyliodonium salts*

Diphenyliodonium salts undergo a very similar photodecomposition pathway producing the same type of photoproducts (Figure 7.3).

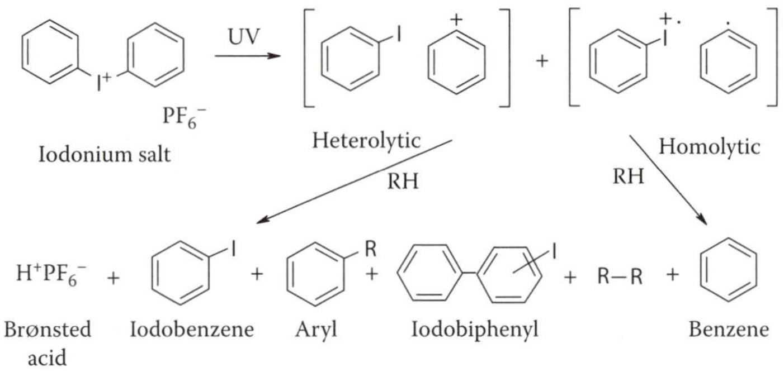

*Figure 7.3* Photodecomposition of an iodonium salt.

Again, the phenyl radical from an unsubstituted aryl group will produce benzene. Iodobenzene is formed and this has its own distinct odor.
Photoproducts of iodonium salts will include isomers of iodobiphenyl.

There is still much to learn about the complex cationic mechanism,
the production of colored species and the polymerization process, and
cationic UV curing is only just beginning to realize its potential and its
advantages over free radical chemistry.

Recent advances are addressing the problems of benzene production
from the photoinitiator and attempts to increase the reactivity are being
studied. New applications are being developed, such as cationic ink-jet
inks, making use of the advantage of the insensitivity to oxygen inhibition
in thin, low-viscosity media. One of the major applications of cationic UV
curing, the polymerization of epoxy silicones, has been studied in some
detail.[6,7] New insights about the reactions of photoinitiators, the epoxy
polymerization process, chain transfer, etc., are certain to appear, and we
are some way from a full understanding of this particular topic.

A cationic photoinitiator should provide good reactivity from its
absorption range, good solubility in the epoxy resins, and good adhesion to
the substrate. In addition, it should not release odorous or toxic by-products.
Recent developments are heading in this direction and new cationic photoinitiators include substituted aryl iodonium salts that do not lead to benzene
production. These salts will, however, release odors related to the substituted aryl structure. A methyl substituted aryl group will release toluene
on exposure, and Irgacure 250 (155), for example, releases isobutylbenzene,
which has an aniseed type of odor. Cyclic sulphonium salts, some based on
the thioxanthone structure, have been made that release very few photoproducts and can be used for low-odor and low-migration applications.

The present range of sulphonium and iodonium salts all suffer from
chromophores that absorb only in the short wave UV. Absorption into the

longer wavelength can be gained for iodonium salts by sensitization with thioxanthones or anthracenes, but sulphonium salts show a much lower response to sensitization. Sensitization can, however, lead to excellent cure in pigmented systems.[8] Another disadvantage is the lack of solubility of these ionic salts in nonpolar formulations. This has been addressed by the use of longer side chain substitution to improve solubility, but in many cases, some reactivity is lost. It is a fact that most forms of substitution on the aryl ring lead to a loss in reactivity compared with the unsubstituted aryl sulphonium or iodonium salt.

## 7.2 Epoxy polymerization and the dark reaction

### 7.2.1 The polymerization process

The Brønsted or Lewis acid that has been formed photochemically from the sulphonium or iodonium salt will open an epoxy group to form a carbonium cation. The cation then attacks another epoxy group and goes on to form a polyether from the ring-opening cationic addition polymerization. The subsequent polymer contains hydroxyl groups and these may be one of the reasons for the excellent adhesion to metals, since they can form hydrogen bonds with any metal oxides on the surface.

All these processes, illustrated in Figure 7.4, are thermally driven and independent of a UV source. The strong Brønsted acid that is originally produced by photochemical reaction is very stable and will continue to promote polymerization long after the UV light is switched off. This "dark reaction" from the acid that remains in the polymer structure leads to full cure over 2–24 hours depending on the temperature, and leads to excellent, uniform conversion of the monomer species. Progression of the dark reaction can be followed by measuring the hardness or the MEK double rubs

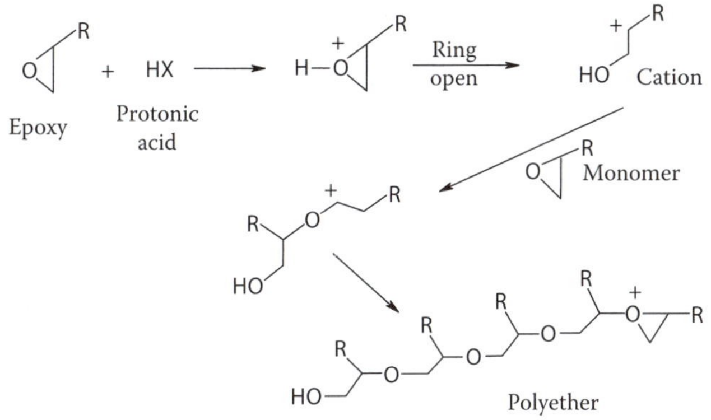

*Figure 7.4* The ring-opening epoxy polymerization.

vs. time, after the initial exposure. Table 7.1 shows the increase in cross-link density.

The cycloaliphatic epoxies are the main resin derivatives used in cationic curing.[9] Epoxy novolak resins are also used for more specific applications. Epoxidized oils such as soya bean, castor and palm oil can be used as diluents. Glycidyl ethers are less reactive than cycloaliphatic epoxies and epoxidized oils.

**Table 7.1** Dark Reaction: Hardness vs. Time

| Time minutes | MEK double rubs |
| --- | --- |
| 15 | 13 |
| 30 | 18 |
| 45 | 24 |
| 60 | 30 |

Vinyl ethers[10] and oxetanes can be used as monomers at 10–20% and will increase cure speed to some degree. Vinyl ethers are very reactive, and multifunctional varieties will produce carbon chain polymers with pendant ethers via a cationic mechanism when used alone. Vinyl ethers can, however, produce aldehyde-type odors in the presence of water. More recently, a variety of oxetanes are coming onto the market and finding more general use as diluents. 2-Ethyl-2-hydroxymethyl oxetane and similar derivatives increase cure speed and hardness and can improve adhesion to difficult substrates. Combinations of oxetanes with glycidyl ethers can offer lower-cost systems that have relatively good cure speeds.

Recent work has indicated that the cationic process is retarded by carbonyl groups including esters. This may be the cause of some conflicting studies on cure speeds where epoxies containing ester groups have reacted more slowly than those with simple alkyl functions. In the same way, polyols that may be included to cross-link and improve cure may have less of an effect if their structures are based on polyol esters.

### 7.2.2   The effect of alcohols and chain transfer

Alcohols will chain transfer from the growing carbonium cation to regenerate acid and effectively terminate the growth of that particular polymerization step[11,12] (Figure 7.5). The acid begins a new polymerization and in effect a multitude of lower molecular weight polymers are produced. This tends to show up as increased cure speed with low levels of alcohol.

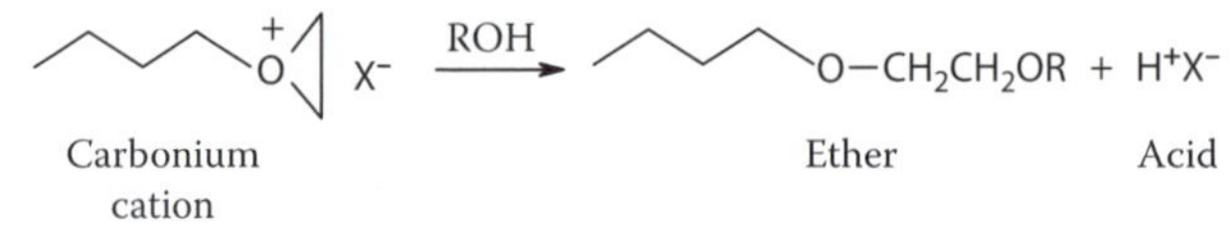

| Carbonium cation | | ROH | Ether | Acid |

*Figure 7.5* Chain termination by an alcohol and regeneration of acid.

The addition of up to 15% tert-butanol has been shown to be more effective than isobutanol or ethanol in increasing the epoxy conversion levels. Diols and polyols will chain extend, increase the flexibility, or cross-link the polymer network when used in small concentrations and have been shown to increase cure speed. Use of 3%–5% of ethylene glycol or butane diol will increase the cure speed of a cationic formulation by 30%–50%. A trifunctional polyol will lead to cross-linking and increase the hardness and the network density of the polymer. However, an excess of polyol, above 20%, may lead to a reduction in cure due to much more extensive chain transfer. Hydroxy substituted epoxy resins have been shown to contribute to adhesive properties.[13]

### 7.2.3  Hybrid cure

Hybrid cationic/free radical curing using a mixture of epoxy and acrylate resins can provide additional advantages for some applications, and the production of superior properties over either system has been claimed.[14] The addition of up to 30% triacrylate resin to the epoxy formulation together with a free radical source such as benzophenone (no amine) or a hydroxyacetophenone can lead to faster cure rates and excellent adhesion properties on plastics and metals. The Type II BP abstracts a hydrogen atom from the polyether or from the substrate, giving an oligomeric radical that can cross-link with the acrylate resin to give a higher density polymer where photografting has occurred. Mono and diacrylate resins are much less effective in hybrid systems.

## 7.3  Triarylsulphonium salts (see Tables A.7 and A.8)

Sulphonium salts show excellent thermal stability, fast surface cure, and good reactivity. Commercial manufacturing processes often produce a mixture of salts (Figure 7.6), including extended conjugation of phenylthio-substituted products. These higher substituted molecules have longer wavelength absorption, higher molar absorption, and are generally more reactive.

The absorption details of the above structures are as follows:

|  | Wavelength max. | Molar absorbance, $lM^{-1}cm^{-1}$ |
| --- | --- | --- |
| Structure 1 | 220–230 nm | 17,500 |
| Structure 2 | 300–310 nm | 19,500 |
| Structure 3 | 300–320 nm | 19,500 |
| Structure 4 | 307–325 nm | 35,500 |

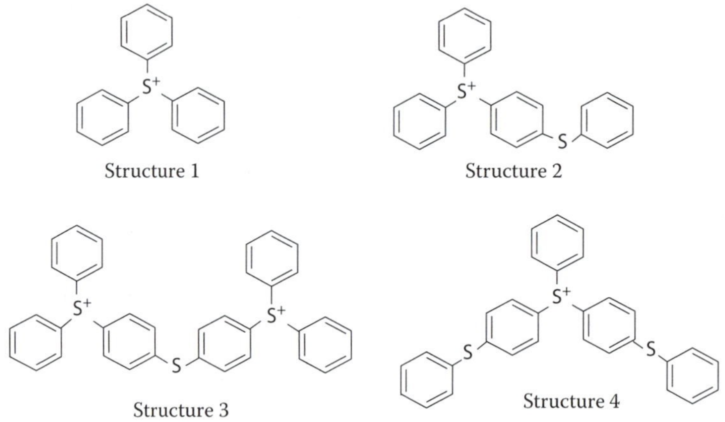

Structure 1                                    Structure 2

Structure 3                                    Structure 4

*Figure 7.6* Extended sulphonium salt structures.

All the structures show a strong UV absorption at 220–230 nm. In addition, structures 2 and 3 have a smaller absorption peak around 300–310 nm, tailing off up to 325 nm, and these show increased reactivity. Structure 4, with its extended absorption and very high molar absorbance, is around three times more reactive[15] than the basic sulphonium salt, Structure 1. This may stem from the ability of these extended compounds to pick up the strong 313 nm output of a mercury lamp. Alternative lamp systems such as the xenon chloride excimer lamp, which gives more intense light at 308 nm, can lead to improved cure of thick coatings but has less effect on surface cure[16] in cationic curing.

The singlet energy $E_s$ of a triphenylsulphonium salt is around 95 kcal/mol and the triplet energy $E_t$ is 75 kcal/mol.

All the above structures with the unsubstituted aryl groups will produce diphenyl sulphide and benzene. Substitution on the aryl ring will counter the production of benzene, and cyclic sulphonium structures will negate the release of diphenyl sulphide. The recent development of novel photoinitiators has addressed these problem areas and led to new materials that are expanding the scope of cationic curing.[17–20]

## 7.3.1   Sulphonium salts that may release benzene

Cyracure UVI-6976 (140) (Dow)

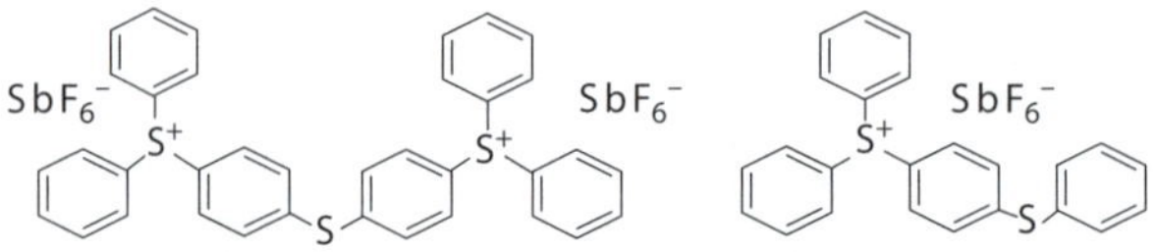

- A mixture of triarylsulphonium salts with the hexafluoroantimonate $SbF_6^-$ anion
  - 50% solution in propylene carbonate
  - Pale yellow liquid
  - Absorption max. ca 300 nm tailing to 330 nm
  - Will produce odor
  - Also marketed as Speedcure 976 (Lambson) and QL Cure 201 (Quang Li Chem.)

Cyracure UVI-6992 (141) (Dow)

- Same mixed salt structure as the previous, with the hexafluorophosphate $PF_6^-$ anion.
  - 50% solution in propylene carbonate.
  - Absorption properties very similar to the previous.
  - A little less reactive.
  - Cyracure 6990 is the same material.
  - Also marketed as Esacure 1064 (Lamberti), Omnicat 432 (IGM), QL Cure 202 (Quang Li Chem.) and Speedcure 992 (Lambson).

Degacure K185 (142). (Degussa)

A bis-sulphonium salt, 33% solution in propylene carbonate.
Absorption up to 320 nm with similar properties to the previous salts.
Also available as Sarcat KI 85 (Sartomer) and SP-55 (Asahi Denko).

## 7.3.2 Sulphonium salts that are "benzene free"

QL Cure 211 (143) (Quang Li Chem.)

- A mixture of sulphonium salts similar to UVI-6976, with p-methyl substituted aryl groups that will release toluene instead of benzene.
  - Hexafluoroantimonate $SbF_6^-$ anion.
  - The performance of any of these substituted cationics is invariably less than that of the unsubstituted, basic sulphonium salt Cyracure UVI-6976.
  - Absorption max. 300–330 nm.

QL Cure 212 (144). (Quang Li Chem.)

- Similar to QL 211 with p-methyl substituted aryl groups and the hexafluorophosphate $PF_6^-$ anion
  - Will produce toluene
  - Slower reactivity than 6992

SP-150 (146). (Asahi Denka)

- The hydroxyethoxy groups improve solubility and are reactive, leading to very low migration.[21]
  - 27% solution in propylene carbonate
  - Absorption 300–320 nm
  - Also marketed as Chivacure R-GEN BF-1172

SP-170 (145). (Asahi Denka)

- Similar to SP-150, with the alternative $SbF_6^-$ anion. A little faster than SP-150.

Omnicat 550 (147) (IGM)

- A sulphonium salt based on the thioxanthone structure.[17] Non-toxic.
  - Low odor and low migration.
  - Does not release diphenyl sulphide.
  - Will release biphenyl, which is non-hazardous.

Omnicat 555 (148). (IGM)

- A solution of Omnicat 550, 40% solids, in monomer.

Omnicat 650 (149) (IGM)

- A polymeric version of Omnicat 550.

Esacure 1187 (150) (Lamberti)

- 9-(4-Hydroxyethoxy)thianthrenium hexafluorophosphate.[19]
  - The structure is based on a cyclic thianthrene.
  - 75% solids in propylene carbonate.
  - Absorption 240 nm and 310 nm.
  - Low odor, low yellowing, low migration.
  - The cure speed is a little slower than the standard sulphonium salt UVI 6992.
  - Can be sensitized to some degree by Anthracure and by Darocur 1173.
  - Shows good radical acceleration with Irgacure 184.

Irgacure MacroCAT (Ciba)

- This is the latest development by Ciba for low migration properties.[23]
  - MacroCAT has a highly substituted aryl structure that delivers low migration, reduced yellowing, and very low odor for food packaging applications.
  - Absorption is extended to 360 nm with improved absorbance.

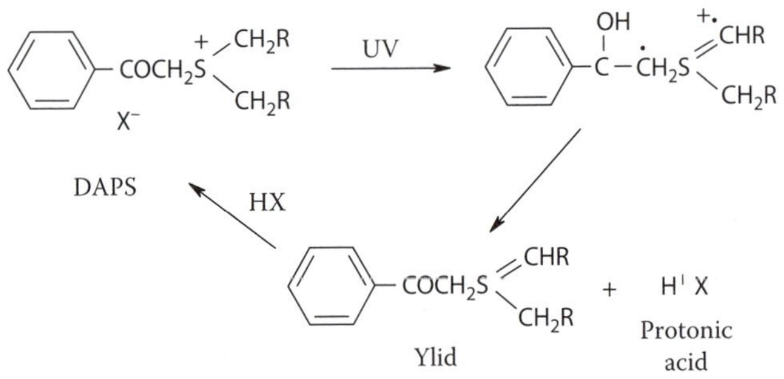

*Figure 7.7* Photodecomposition of dialkylphenacylsulphonium salts.

- Reactivity is a little less than the standard sulphonium salt but it can be improved with the addition of small amounts of DBA sensitizer.

## 7.4   Dialkylphenacylsulphonium salts

Dialkylphenacylsulphonium salts (DAPS)[1,23,24] have been studied academically in some detail and show excellent reactivity.

When irradiated, the dialkylphenacyl salt shown in Figure 7.7 undergoes an internal hydrogen abstraction mechanism to produce a radical cation-radical pair. Electron transfer and rearrangement produces a proton and an ylid. The ylid can recombine with the acid in a dark reaction to regenerate the original salt. This means that the acid is produced under UV irradiation as usual, but when the irradiation stops, the acid is removed by reaction with the ylid. There is no residual acid left in the system to sustain the normal dark reaction, and the polymerization becomes purely photosensitive, which may be an advantage to some applications.

The first-generation simple alkyl phenacyl salts have a poor solubility profile, but this can be greatly improved by extended alkyl groups. These second-generation materials show low odor due to the few photoproducts that are produced and can be readily sensitized in the long wave UV by electron transfer sensitizers. They do not release benzene. The phenacyl salts become bound to the polymer and show very low migration. As yet, there are no commercial sources of these salts.

## 7.5   Diaryliodonium salts

Diaryliodonium salts[1] (see Tables A.7 and A.8) are less reactive and less thermally stable than sulphonium salts, and the simpler salts show some degree of toxicity. The low reactivity may be due to the very short wave UV chromophore at 220–270 nm, which is unable to absorb the higher UV

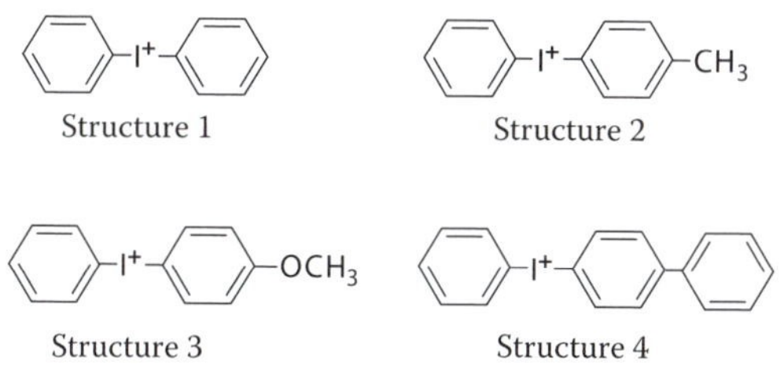

Structure 1

Structure 2

Structure 3

Structure 4

*Figure 7.8* Substituted iodonium salts.

*Table 7.2* Wavelength and Absorbance
of Iodonium Salts

|  | Wavelength nm | Molar absorbance 1.M$^{-1}$ cm$^{-1}$ |
|---|---|---|
| Structure 1 | 227 | 18,000 |
| Structure 2 | 237 | 18,000 |
| Structure 3 | 246 | 18,500 |
| Structure 4 | 267 | 21,000 |

energies that are available at longer wavelengths (Figure 7.8). Absorption and reactivity increase with electron donating substitution on the ring, as for the sulphonium salts, but with less effect (Table 7.2).

The iodonium salts have an absorption that tails off at about 300 nm and are unable to pick up the 313 nm output line of a mercury lamp. They can respond only to the weaker 254 nm output and on their own are very poor photoinitiators. Alkoxy substitution increases the cure speed a little[25] and long chain alkyl substitution, such as dodecyl,[26,27] increases the solubility and reduces the inherent toxic nature of these salts. The triplet energies of iodonium salts are around 64 kcal/mol.

The reactivity of iodonium salts is much less than that of sulphonium salts. Long wave sensitization by a thioxanthone and radical acceleration will increase the reactivity of the iodonium salts[28] to at least that of the sulphonium salts, if not better, and lead to excellent depth cure in thick films or pigmented media.

Iodonium salt cure speed is at best 50–60% that of sulphonium salts. Iodonium salt plus anthracene or thioxanthone sensitizer, and iodonium salt plus 1173 radical generator gives similar cure to sulphonium salts. Iodonium salts can excel for depth cure when used with thioxanthone long wave sensitizers. For example, 1.5% iodonium salt plus 0.5% ITX or CPTX will give excellent depth cure.

For both types of cationic salt, the solubility of the ionic salt in the non-polar medium often has a bearing on the reactivity of the formulation.

## 7.5.1   Iodonium salts that may produce benzene

Hycure-810 (151) (ChemFine)

- Diphenyliodonium hexafluorophosphate
  - The basic iodonium salt
  - Absorption 230–260 nm
  - Poor solubility

Uvacure 1600 (152) (Surface Specialties [UCB])

- Phenyl-p-octyloxyphenyl iodonium hexafluoroantimonate
  - MP 57–58°C
  - Absorption 240 nm
  - Much improved solubility and reduced toxicity

Sarcat CD-1012 (153) (Sartomer)

- Similar to the above, with improved solubility and reduced toxicity
  - Absorption 240 nm

## 7.5.2   Iodonium salts that are "benzene free"

Omnicat 440 (154) (IGM)

- Bis-(4-methylphenyl)iodonium hexafluorophosphate
  - White powder
  - MP 175–180°C
  - Absorption max. 267 nm

- Can be sensitized with CPTX, etc.
- Releases toluene
- Also marketed as Hycure-820 (ChemFine)

Omnicat 445 (IGM)

- A 50% solution of Omnicat 440 in 3-ethyl-3-hydroxymethyl oxetane.

Irgacure 250 (155) (Ciba)

- 4-Isobutylphenyl-4′-methylphenyliodonium hexafluorophosphate.[29–31]

- Produces isobutylbenzene on photolysis
  - 75% solution in propylene carbonate
  - Can be efficiently sensitized with CPTX, ITX or Anthracure
  - Use 2–3% salt plus 0.5–1% CPTX to respond to long wave UV

UV 9310 (156) (GE)

- Bis-(4-dodecylphenyl)iodonium hexafluoroantimonate
  - The long alkyl chains reduce toxicity and lead to excellent solubility[25] due to the multiple isomer nature of the product. The reactivity is reduced due to the higher molecular weight.

Rhodorsil 2074 (157) (Rhodia)

- Tolylcumyliodonium tetrakis(pentafluorophenyl) borate[32–35]
  - White powder
  - MP 120–133°C
  - Abs. range 230–300 nm
  - Releases cumene (isopropylbenzene)
  - Also available as 18% solids in isopropanol

- Extremely fast cure in epoxy silicones but less advantageous in cycloaliphatic epoxies
- Can be sensitized by thioxanthones and anthracenes

Rhodorsil 2076 (158) (Rhodia)

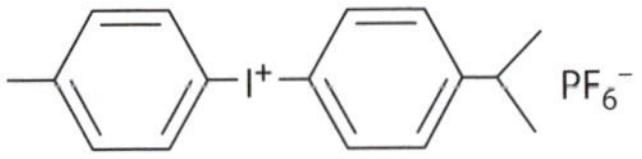

- 4-Isopropylphenyl-4'-methylphenyliodonium hexafluorophosphate 50% solution in propylene carbonate

## 7.6   Ferrocenium salts

Ferrocenium salts are very stable salts that can dissociate under UV energy to produce Lewis acids capable of epoxy polymerization.[36] (See Figure 7.9.)

The ferrocenium salt loses cumene under UV and forms a Lewis acid ligand with the epoxy groups. This type of acid is less reactive than a Brønsted acid and requires a thermal step following UV exposure to complete the polymerization. Ferrocenium salts photobleach, giving excellent depth cure, and can be used for very thick film applications. They are used for high contrast imaging with epoxy novolaks, but tend to suffer from poor solubility.

Irgacure 261 (159) (Ciba)

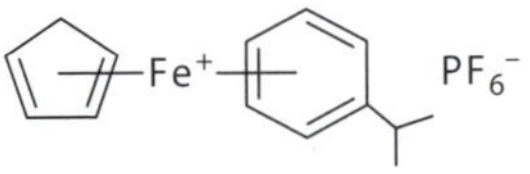

- Cyclopentadienylcumene-iron hexafluorophosphate
- Has a lower reactivity than the onium salts and requires a thermal step to complete the cure
- Absorption maximum 240 nm
- 75% solids in solution

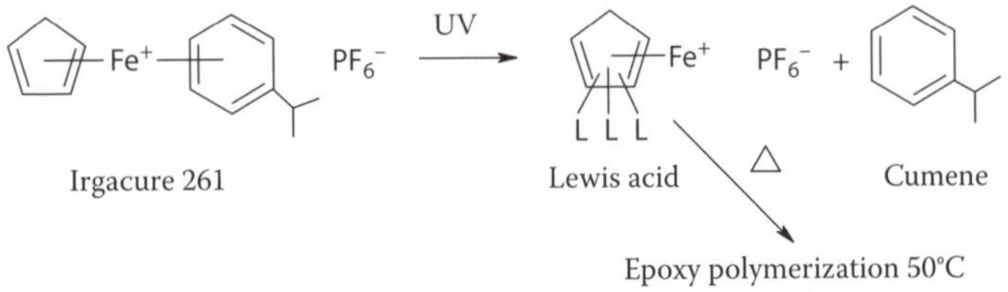

*Figure 7.9* Photodecomposition of a ferrocenium salt.

## References and further reading

1. Crivello, J. V. (Rensselaer Poytechnic Inst. NY.) Recent progress in cationic photopolymerization. *Proc. PRA Formulation and Performance Conference,* Harrogate (UK), 2000, Paper 10.
2. Hofer, M., C. Heller, R. Liska. (Vienna Univ.) New sulphonium salt based photoinitiators for cationic photopolymerization. *RadTech Eu. Conf. Proc.* 2007. Posters—Advances in photochemistry and polymerization.
3. Crivello. J. V. *Adv. Polym. Sci.* Cationic polymerisation. Iodonium and sulphonium salt photoinitiators. 62 (1984), 1. (Review.)
4. Holman, R. 2000. Cationic systems are not at a crossroads. *PRA RADnews* Spring, 32.
5. Carter, J. W., M. J. Jupina. (Union Carbide) Cationic UV ink migration and safety assessment. *RadTech Eu. Conf. Proc.* 1997. 250–258.
6. Priou, C. (Rhone Poulenc.) Cationic photoinitiators for curing epoxy-modified silicones. *Proc. PRA Conference.* Harrogate (UK), 1996. Paper 14, 2–15.
7. Priou, C., J. M. Francis, J. Richard, S. Kerr. (Rhodia Silicones.) UV/EB curing of epoxy silicone coatings. *PRA RADnews.* Summer 1998, 25, pp. 9–14.
8. Carroy, A. (Union Carbide). New developments in the formulation of cationic UV curing systems. *RadTech Eu. Conf. Proc.* 1995. 523.
9. Carroy, A., F. Chomienne. (Union Carbide) Cationic UV curing of clear and pigmented system based on cycloaliphatic epoxide/vinyl ether blending. *RadTech NA. Conf. Proc.* 2000. 10–21.
10. Abadie. M. J. M., L. Ionescu-Vash. (Montpelier Univ.) Recent advances in cationic photoinitiators. *RadTech Asia. Conf. Proc.* 1995. 52–58.
11. M. Sangermano. (Polytecnico di Torino) New developments in cationic photopolymerization for coatings applications. *Proc. PRA Economy and Performance Conference.* Manchester (UK), 2004, Paper 1.
12. J. V. Crivello, D. A. Conlon, D. R. Olson, K. K. Webb. (General Electric) Accelerators in UV cationic polymerization. *RadTech Eu. Conf. Proc.* 1987. 1–27 to 1–39.
13. A. Carroy. (Union Carbide) Cationic radiation cured coatings: Physico-chemical properties related to formulation and irradiation. *RadTech Eu. Conf. Proc.* 1991. Paper 23, 265–283.
14. R. S. Davidson. *PRA RADnews,* (Hybrid cure) 2008, 65, pp. 34–44.
15. J. V.Crivello, J. L. Lee, D. A. Conlon, Makromol. Chem. *Macromol. Symp.* 13/14 (1988), 145.
16. A. Carroy, F. Chomienne, J. F. Nebout. (Union Carbide). Recent progress in the photoinitiated cationic polymerization of cycloaliphatic epoxy systems. *RadTech Eu. Conf. Proc.* 1997. 303–313.
17. Herlihy. S., B. Rowatt, R. S. Davidson. (Sun Chemical) Novel sulphonium salt cationic photoinitiators for food packaging applications. *RadTech Eu. Conf. Proc.* 2003. 225–232.
18. Davidson, R. S. (Citifluor Ltd.) Whither cationic curing? The development of new photoinitiators. *Proc. PRA Economy and Performance Conference.* Manchester (UK), 2004, Paper 2.
19. Casiraghi, A., M. Cattaneo, G. Norcini, M. Visconti. (Lamberti). Novel cationic photoinitiators. *RadTech NA. Conf. Proc.* 2004, 2,6.

20. Visconti. M., A. Casiraghi. E. Bellotti. (Lamberti) A novel cationic photoinitiator. *RadTech Eu. Conf. Proc.* 2005. 271–278.
21. Tsuchiya, H., K. Morio, H. Murase, K. Ohkawa. 1987. *USP* 4,684,771.
22. Crivello, J. V., J. H. W. Lam. 1979. Diakylphenacyl sulphonium salts. *J. Poly. Sci. Polym. Chem.* 17, 2877.
23. Studer, K. et al. (Ciba/Mirage) Design of high molecular weight photoinitiators for high end UV curing applications. *RadTech Eu. Conf. Proc.* 2009.
24. Crivello, J. V., J. L. Lee. Structure and mechanistic studies in the photolysis of dialkylphenacyl sulphonium salt cationic photoinitiators. *Macromolecules.* 16 (1983), 864.
25. Crivello, J. V., J. L. Lee. (Rensselaer Polytechnic Inst. NY.) Alkoxy substituted diaryliodonium salt cationic photoinitiators. *RadTech NA. Conf. Proc.* 1990. 424–431.
26. Eckberg, R., S. Rubinsztajn. (General Electric) Long chain iodonium salts. *RadTech NA. Conf. Proc.* 2002. p 41.
27. Eckberg, R., K. D. Riding. In *Radiation Curing of Polymeric Materials*. Eds. C. E. Hoyle and J. F. Kunstle. Washington, D.C.: ACS. 392.
28. Caiger, N. A., D. H. Selman, G. E. C. Beats. PCT Intnl. Patent Appln. WO 017644, 2007.
29. Birbaum, J-L., S. Ilg. (Ciba) A new onium salt for cationic curing of white pigmented epoxy formulations. *RadTech Europe 2001.*
30. Carroy, A., F. Chomienne, J. F. Nebout. (Dow Europe) Advances in cationic curing of cycloepoxide systems. *RadTech Eu. Conf. Proc. 2001.*
31. Birbaum, J-L., S. Ilg, T. Bolle, E.V. Sitzmann, D.A. Wostratzky, A. Carroy. (Ciba) A novel, benzene free iodonium salt. *RadTech NA. Conf. Proc.* 2002. p. 28.
32. Priou, C., J. M. Frances, J. Richards, S. Kerr. (Rhodia Silicones). 1998. UV/EB curing of epoxy silicone coatings. *PRA RADnews* Summer 25, 9–14.
33. Priou, C., J. M. Frances. (Rhodia Silicones) New developments in the field of cationic photocrosslinking of epoxy resins with borate photoinitiators. *RadTech NA. Conf. Proc.* 1998. 476–485.
34. Priou, C. (Rhone-Poulenc) Cationic photoinitiators for curing epoxy modified silicones. *Proc. PRA Conference.* Harrogate (UK), 1996, Paper 14.
35. Feng, K., H. Zang, D. C. Neckers. (Bowling Green State Univ.) D. Martin, T. L. Marino. (Spectra Group Ltd) Synthesis and study of iodonium borate salts as photoinitiators. *RadTech NA. Conf. Proc.* 1998, 215–227.
36. Meier, K. H. Zweifel. *Radiation Curing Conf. Basel.* 1985. SME, Paper FC85–417.

# chapter eight

# Factors affecting the use of cationic photoinitiators

## 8.1   The influence of the anion

The non-nucleophilic anions that accompany these cationic sulphonium and iodonium salts have no influence on the photodecomposition of the cationic photoinitiator; the decomposition and the spectral response depend purely on the structure of the aromatic groups on the onium cation.

However, the strength of the acid produced depends inversely on the nucleophilicity of the anion and this is of great importance when it comes to the thermal polymerization process and the ring opening mechanism.

In general, the reactivity of the anion increases:

$$BF_4^- < PF_6^- < AsF_6^- < SbF_6^- \ldots\ldots < B(C_6F_5)_4^-$$

The hexafluoroantimonate anion, $SbF_6^-$, is by far the most reactive of the more common anions, with the tetrafluoroborate anion being very slow curing in comparison. In terms of conversion, the $PF_6^-$ anion may give around 70% compared with that of the $SbF_6^-$ anion, but cure speeds will also vary with many other factors including the type of media used, reactive diluents, temperature, etc. The toxic nature of the antimony and arsenic salts in these anions means that present developments in onium salts are mostly confined to the hexafluorophosphate anion $PF_6^-$ on environmental grounds.

The tetrakis-(pentafluorophenyl)-borate anion was developed primarily for use in silicone-based epoxy resin systems for release papers, etc. This anion has an extremely low nucleophilicity due to the very bulky structure and the charge separation it produces. In an epoxy silicone environment it is the by far the most reactive of all the anions, being about ten times faster than the antimonate anion.

Unlike the inorganic anions, the aryl structure of this complex borate anion has some effect on the UV absorption properties at 200–240 nm. At 220 nm the molar absorption of the borate anion is 40,000 $l.M^{-1}\ cm^{-1}$, which is twice that of the simple cationic $PF_6^-$ salt (18,000 $l.M^{-1}\ cm^{-1}$), and this may be one factor in its excellent cure speed. However, this borate anion is less advantageous in the more common cycloaliphatic epoxy resin formulations and generally performs little better than the antimonate salt.

Other anions, such as tosylates, triflates, and perchlorates, are insufficiently active to be used for ring opening epoxy polymerizations but may find use as photolatent acids in imaging, color formation, and acid deblocking applications.

## 8.2 Photochemical radical decomposition of onium salts

Free radicals can promote the generation of cations by a redox mechanism with an onium salt.[1] Figure 8.1 and Figure 8.2 illustrate this process.

The carbon-centered cation produced from the radical may initiate the ring-opening polymerization or go on to produce a proton.

Alpha hydroxy radicals (from a hydroxyacetophenone photoinitiator) and alpha ether radicals (by hydrogen abstraction from the oligomer) have strong reducing properties and can accelerate the photodecomposition of the cationic salt. The hydroxyalkyl radical in Figure 8.2 undergoes electron transfer with the cation to give acetone and produce a proton. The most efficient photoinitiators for this use are the Type I hydroxyacetophenones such as Darocur 1173 (1), Irgacure 184 (3), Irgacure 2959 (4) and Irgacure 127 (7).

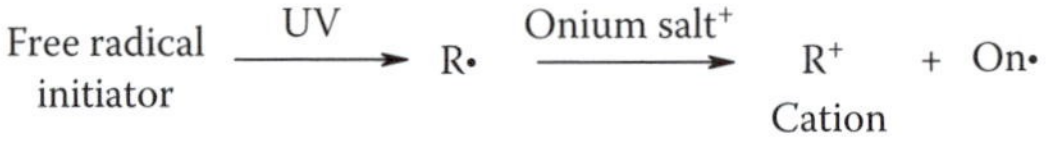

*Figure 8.1* Cation generation from a radical source.

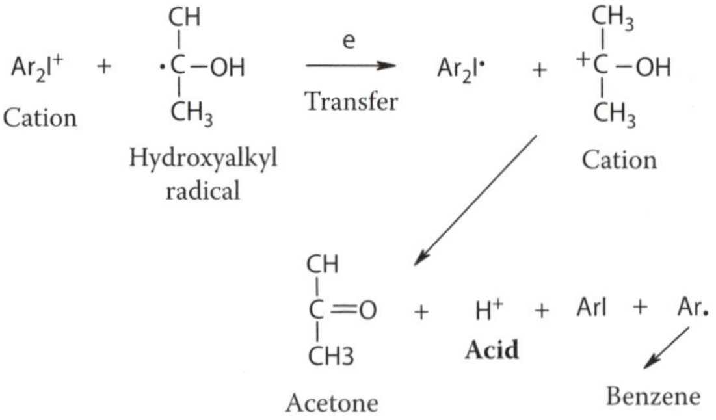

*Figure 8.2* Radical acceleration of the decomposition of an iodonium salt.

The thioxanthone ketyl radical is also effective (by hydrogen atom abstraction from the oligomeric ether) and is more reactive than that of benzophenone in the Type II series. Amines must not be added if Type II initiators are used. Recent work has indicated that radicals from the titanocene photoinitiator, Irgacure 784 (42) absorbing in the visible at 438 nm, will readily reduce an iodonium salt and promote cationic polymerization.[2] The phosphinyl radicals from Lucirin TPO (22) and Irgacure 819 (24) are also effective "radical accelerators."[3] Iodonium salts are much more responsive to radical decomposition than sulphonium salts.

## 8.3   Sensitization of the cationic photoinitiator

### 8.3.1   Extension of absorption wavelength by a sensitizer

Most cationic photoinitiators absorb in the short wave UV up to 300 nm and are generally inefficient when used in pigmented media. Sensitization by materials absorbing in the long wave UV would allow better use of the available light energy. Sensitization in free radical systems is mostly via energy transfer, but for cationic photoinitiators, which have very high singlet and triplet energies, this is not possible. Sensitization can occur by electron transfer, as shown in Figure 8.3, based on the relative redox potentials of the sensitizer and cation. The sensitizer, unlike in a free radical system, where it is catalytic, is involved in the reaction and becomes arylated.

Sensitizers that can be used with iodonium salts[4] include dialkoxyanthracene[5] derivatives such as 2-ethyl-9,10-dimethoxyanthracene (EDMA) (162), 9,10-diethoxy (160) and 9,10-dibutoxyanthracene (161) (the Anthracures, DEA and DBA), and thioxanthones[6,7] such as ITX (63) and CPTX (69).

Sensitization of sulphonium salts is much less efficient[8,9] due to their greater redox potential, and only the Anthracures® with singlet energies $E_s$ of 76 kcal/mol will offer some improvement in this case.

Typical concentrations used for radical accelerators and sensitizers are 3% cationic salt, 1% hydroxyalkylacetophenone and 0.5% thioxanthone.

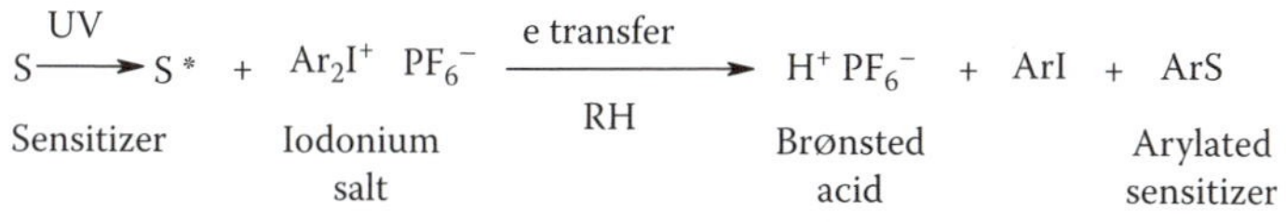

$$S \xrightarrow{\text{UV}} S^* \; + \; Ar_2I^+ \; PF_6^- \; \xrightarrow[\text{RH}]{\text{e transfer}} \; H^+ \, PF_6^- \; + \; ArI \; + \; ArS$$

| Sensitizer | Iodonium salt | | Brønsted acid | Arylated sensitizer |

*Figure 8.3*  Long wave sensitization of a cationic salt.

*Table 8.1* Sensitizers for Cationic Salts

| Sensitizer | Absorption nm | Sulphonium salt | Iodonium salt |
|---|---|---|---|
| ITX | 383 | No | Yes |
| CPTX | 387 | No | Yes |
| Anthracure UVS-1101 (DEA) | 402 | Yes | Yes |
| Anthracure UVS-1331 (DBA) | 403 | Yes | Yes |
| EDMA | 401 | Yes | Yes |
| Benzophenone | 254 | No | Yes |
| Acetophenone | 278 | Yes | Yes |
| Acetone | 279 | Yes | Yes |
| Perylene | 410 | Yes | Yes |
| Irgacure 784 | 460 | No | Yes |

### 8.3.2   Sensitizers for cationic photoinitiators

Some of the more common sensitizers are listed in Table 8.1.

EDMA, 2-ethyl-9,10-dimethoxyanthracene, is a laboratory chemical, not yet available commercially. Benzophenone, acetophenone,[10] and acetone are short wave UV sensitizers that can boost the cure of clear cationic systems. Perylene, absorbing at 388 nm, 410 nm, and 437 nm, and similar polyaromatics have been used academically as sensitizers.

## 8.4   The influence of temperature on the polymerization

The initial step in the decomposition of a cationic salt is photochemically controlled, leading to a stable Brønsted acid, but the following ring-opening mechanism and the polymerization process, including the dark reaction, are all thermally sensitive. The temperature of the curing process and the post cure therefore has a much greater effect on polymerization rates and final cure properties than those of free radical systems.[11,12] Typical cure speeds vs. temperature are illustrated in Figure 8.4.

UV curing at higher temperatures, approaching 100°C, will lead to very fast cure. Alternatively, a thermal post-cure step at 60°C will increase hardness levels by 30% against those at ambient temprature, and accelerate full cure via the dark reaction of the Brønsted acid to almost 100% conversion.

## 8.5   The effect of water on polymerization

Water can act as a chain transfer agent in cationic curing, promoting new chain growth, reducing the average molecular weight, slowing cure and reducing hardness.[13,14]

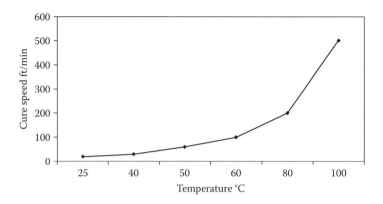

*Figure 8.4*  Cure speed vs. temperature for cationic polymerization.

To generate protons, a small amount of water can interact with the hydrogen acid that is formed, possibly accelerating cure, but at higher concentrations chain transfer dominates. Water in the bulk of the formulation up to around 5% has been shown to have little influence on chain propagation, but atmospheric water in the form of humidity has a more significant effect (see Figure 8.5) and will cause chain transfer, destroying much of the film properties. Permeability of water throughout the surface of the coating has more influence and an increase in the thickness of the coating will reduce these effects.

The effect of water on cationic cure depends to some extent on the formulation and structure of the epoxide used. Using formulations with a high hydrocarbon content or adding a small amount of silicone derivative will give a hydrophobic mixture that is less susceptible to the effects of water. Vinyl ethers are particularly prone to these effects. Under normal

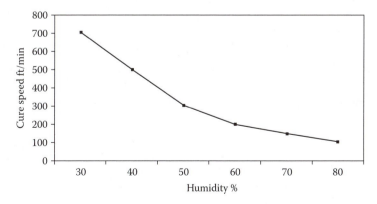

*Figure 8.5*  Cure speed vs. humidity for cationic polymerization.

conditions, the presence of water is a surface phenomenon predominantly influenced by high relative humidity, which reduces the cure speed. Humidity can also have an effect on the post cure, hindering full conversion since the dark reaction can extend for a significant length of time.[15–17]

Increasing the temperature at the air–coating interface by the use of infrared or hot air blowers prior to cure reduces the effects of humidity and can make a significant improvement in cure. Alternatively, raising the substrate temperature to around 50°C during cure will nullify most of these humidity effects.

## References and further reading

1. Baumann, H., H. J. Timpe. 1984. *J. Prakt. Chem.* 326, 529.
2. Degirmenci, M., A. Onen, Y. Yagci. Visible light photoinitiation of cationic polymerization by using titanocene in the presence of onium salts. Istanbul Technical University, Maslak, Istanbul 80626, Turkey.
3. Dursun, C., M. Degirmenci, Y. Yagci, S. Jockush, N. J. Turro. 2003. Free radical promoted cationic polymerisation by using bisacylphosphine oxide photoinitiators: Substituents effect on the reactivity of phosphenoyl radicals. *Polymer* 44, 7389–7396.
4. Devoe, R. J., M. R. V. Sahyun, N. Serpone, D. K. Sharma. 1987. *Can. J. Chem.* 65, 2342.
5. Crivello, J. V., M. Jang. (Rensselaer Polytechnic Inst. NY). 2003. Anthracene electron transfer photosensitizers for onium salt induced cationic photopolymerization. *J. Photochem. Photobiol. A: Chem.* 159, 173–188. www.sciencedirect.com.
6. Crivello, J. V., J. L. Lee. (Rensselaer Polytechnic Inst. Troy, NY) Alkoxy substituted diaryliodonium salt cationic photoinitiators. *RadTech NA. Conf. Proc.* 1990. 424–431.
7. Birbaum, J-L., S. Ilg. (Ciba) A new onium salt for cationic curing of white pigmented epoxy formulations. *RadTech Eu. Conf. Proc.* 2001. 545–550.
8. Dektar, J. L. and N. P. Hacker. 1988. Triphenylsulphonium salt photochemistry. New evidence for triplet excited state. *J. Org. Chem.* 53, 1833.
9. Dektar, J. L., N. P. Hacker, K. M. Welsh, and N. J. Turro. 1989. Photo-CIDNP and nanosecond flash photolysis studies on the photodecomposition of triarylsulphonium and diarylhalonium salts. *Polym. Mater. Sci. Eng.* 6, 181.
10. Dektar, J. L. and N. P. Hacker. 1990. Photochemistry of diaryliodonium salts. *J. Org. Chem.* (55), 639.
11. Crivello, J. V., D. A. Conlon, D. R. Olson, K. K. Webb. (General Electric) Accelerators in UV cationic polymerization. *RadTech Eu. Conf. Proc.* 1987. 1–27 to 1–39.
12. Popp, M., A. Harwig, and K. Teczyk. (Fraunhofer Institute, Germany). Influence of temperature and atmosphere on the curing rate and final properties of UV curing acrylate and epoxides. *RadTech Eu. Conf. Proc.* 607–614.
13. Brann, W. L.. (Lord Corp.) The effects of moisture on UV curable cationic epoxide sustems. *RadTech Eu. Conf. Proc.* 1989. 565–579.

14. Hupfield, P. C., S. R. Hurford, J. S. Tonge. (Dow Corning) The effect of moisture on the cationic polymerization. *RadTech NA. Conf. Proc.* 1998. 468–475.
15. Paul, C. W., C. Meisner, and P. Walter. Cationic UV-curable hot melt pressure sensitive adhesives. *RadTech NA. Conf. Proc.* 2.5.
16. Davidson, R. S., K. S. Tranter. (City Univ. UK) Use of remote curing for probing the effects of humidity on the cationic curing of epoxides. *RadTech Eu. Conf. Proc.* 2007. Application Properties.
17. Z. Cheng. (Centre for Nanoscale Sc. & Eng, NDSU) A real-time FTIR study of the humidity effect on cationic photopolymerization of epoxy-siloxane composites. *RadTech NA. Conf. Proc.* 2008.

# appendix A

## Tables and absorbance graphs

Trademarks can be bought out and transferred. Suppliers change
and merge and new agencies spring up all the time, so this list will
inevitably be incomplete.

Includes trade name, risk phrases, and registration numbers where
available. Registrations in some countries can change as more data
is assembled. The list will need updating.

*Table A.1* Type I Photoinitiators: Physico-Chemical Data

| No. | Trade name abbreviation | Structure Chemical name/Synonym | Molecular formula Mol. Wt. | Form MP/BP °C |
|---|---|---|---|---|
| 1 | **Hydroxy-acetophenones** Additol HDMAP Chivacure 173 Darocur 1173 Esacure KL200 Firstcure HMPP Genocure DMHA Hycure 1173 Omnirad 73 Runtecure 1103 SarCure SR 1121 Speedcure 73 | 2-hydroxy-2-methyl-propiophenone 2,2-dimethyl-2-hydroxyacetophenone 2-hydroxy-2-methyl-1-phenylpropanone | $C_{10} H_{12} O_2$ 164.2 | liquid MP 4 BP 80–81 |
| 2 | Hycure 185 Omnirad 102 | 2-hydroxy-2-methyl-4'-tert-butyl-propiophenone 2-hydroxy-2-methyl-1-(4-tert.butylphenyl)-propanone | $C_{14} H_{20} O_2$ 220.4 | liquid 98–100 4 mm |

*(continued on next page)*

*Table A.1 (continued)* Type I Photoinitiators: Physico-Chemical Data

| No. | Trade name abbreviation | Structure Chemical name/Synonym | Molecular formula Mol. Wt. | Form MP/BP °C |
|-----|------------------------|--------------------------------|----------------------------|---------------|
| 3 | Additol CPK<br>Chivacure 184<br>Esacure KS300<br>Firstcure HCPK<br>Genocure CPK<br>Hycure 184<br>Irgacure 184<br>Omnirad 481<br>PI 184<br>Runtecure 1104<br>SarCure SR 1122<br>Speedcure 84 | 1-hydroxycyclohexyl phenyl ketone | $C_{13} H_{16} O_2$<br><br>204.3 | solid<br><br>47–50 |
| 4 | Irgacure 2959<br>IGM 2959<br>Hycure 2959<br>PI-2959 | 2-hydroxy-4'-(2-hydroxyethoxy)-2-methyl-propiophenone | $C_{12} H_{16} O_4$<br><br>224.3 | Solid<br><br>88–90 |

| | | Structure | Chemical name | Formula | MW | Form | mp |
|---|---|---|---|---|---|---|---|
| 5 | Omnirad 669 | | 2-hydroxy-[4'-(2-hydroxypropoxy)]-2-methyl-propiophenone<br>2-hydroxy-[4'-(2-hydroxypropoxy)phenyl]-2-methylpropanone | $C_{13}H_{18}O_4$ | 238.3 | solid | 76–79 |
| 6 | Esacure KIP 150 polymeric<br>KIP EM (emulsion) | | oligo 2-hydroxy-2-methyl-1-[4-(1-methyl-vinyl)phenyl]propanone | $(C_{13}H_{16}O_2)n$<br>ca 800 | | gum | |

*(continued on next page)*

Table A.1 (continued) Type I Photoinitiators: Physico-Chemical Data

| No. | Trade name abbreviation | Structure Chemical name/Synonym | Molecular formula Mol. Wt. | Form MP/BP °C |
|---|---|---|---|---|
| 7 | Irgacure 127 | <br>2-hydroxy-1-[4[4-(2-hydroxy-2-methyl-propionyl)-benzyl]-phenyl]-2-methyl-propan-1-one | $C_{21} H_{24} O_4$<br>340.4 | solid<br>82–90 |
| 8 | Esacure ONE | <br>mixed isomers | $C_{26} H_{33} O_4$<br>408 | |
| 9 | Irgacure PICS-1 | Dendrimeric hydroxyacetophenone | | |

| | | | $C_{15}H_{21}NO_2S$ 279.4 | solid 74–76 |
|---|---|---|---|---|
| 10 | **Alkylamino-acetophenones** Additol DMTTA Chivacure 107 Genocure PMP Hycure 907 Irgacure 907 Irgacure 907FF Nanocure 907 Omnirad 4817 Runtecure 1107 Speedcure 97 | 2-methyl-4′-(methylthio)-2-morpholino-propiophenone | | |
| 11 | Quadracure MMMP-3 | R is e-caprolactone oligomer | ca. 630 | liquid |

(continued on next page)

*Table A.1 (continued)* Type I Photoinitiators: Physico-Chemical Data

| No. | Trade name abbreviation | Structure Chemical name/Synonym | Molecular formula Mol. Wt. | Form MP/BP °C |
|-----|-------------------------|----------------------------------|-----------------------------|----------------|
| 12 | Irgacure 369 PI-369, Genocure BDMM, Speedcure BDMB | 2-benzyl-2-(dimethylamino)-4-morpholinobutyrophenone | $C_{23}H_{30}N_2$ $O_2$ 366.5 | solid 110–114 |
| 13 | Quadracure BDMD-3 | R is e-caprolactone oligomer | ca. 720 | liquid |

| 14 | Irgacure 379 | 2-(4-methylbenzyl)-2-(dimethylamino)-4-morpholinobutyrophenone | $C_{24} H_{32} N_2$ $O_2$ 380.5 | solid 82–87 |
| 15 | Omnipol 910 polymeric | polymeric aminoalkylphenone | $C_{62} H_{88}$ $N_4 O_{10}$ avg. 1039 | liquid polymer |

(continued on next page)

Table A.1 (continued) Type I Photoinitiators: Physico-Chemical Data

| No. | Trade name abbreviation | Structure Chemical name/Synonym | Molecular formula Mol. Wt. | Form MP/BP °C |
|---|---|---|---|---|
| 16 | **Benzil ketals, etc.** Additol BDK Chivacure BDK Esacure KB1 Firstcure BDK Genocure BDK Hycure BDK Irgacure 651 Omnirad BDK Micure BK6 Nanocure BDK PI 651 Runtecure 1065 SarCure SR 1120 Speedcure BKL | 2,2-dimethoxy-2-phenyl-acetophenone benzil dimethyl ketal | $C_{16} H_{16} O_3$ 256.3 | solid 64–67 |
| 17 | Uvatone 8302 | 2,2-diethoxy-2-phenylacetophenone | $C_{18} H_{20} O_3$ 284.4 | solid |

| No. | Name | Structure / chemical name | Formula / MW | State / bp |
|---|---|---|---|---|
| 18 | DEAP<br>Firstcure<br>IGM<br>Rahn | 2,2-diethoxy-1-phenylethanone<br>diethoxyacetophenone | $C_{12}H_{16}O_3$<br>208.3 | liquid<br>131–134<br>10 mm |
| | **Benzoin ethers** | | | |
| 19 | BIPE<br>Daitocure IP<br>Vicure 30 | 2-isopropoxy-2-phenylacetophenone | $C_{17}H_{18}O_2$<br>254.3 | solid |
| 20 | BNBE<br>Esacure EB1<br>Trigonal 14 | 2-n-butoxy-2-phenylacetophenone | $C_{18}H_{20}O_2$<br>268.4 | solid |
| 21 | BIBE<br>Daitocure IB<br>Esacure EB3<br>Vicure 10 | 2-isobutoxy-2-phenylacetophenone | $C_{18}H_{20}O_2$<br>268.4 | liquid<br>133/0.5 mm |

(continued on next page)

*Table A.1 (continued)* Type I Photoinitiators: Physico-Chemical Data

| No. | Trade name abbreviation | Structure Chemical name/Synonym | Molecular formula Mol. Wt. | Form MP/BP °C |
|---|---|---|---|---|
| | **Phosphine oxides** | | | |
| 22 | Additol TPO | | $C_{22}H_{21}O_2P$ | solid |
| | Chivacure TPO | | | 87–94 |
| | Darocur TPO | | 348.4 | |
| | Esacure TPO | | | |
| | Firstcure TPO | | | |
| | Genocure TPO | | | |
| | Goldcure TPO | | | |
| | Hycure TPO | Diphenyl-(2,4,6-trimethylbenzoyl)- | | |
| | Lucirin TPO | phosphine oxide | | |
| | Lucirin LR 8953X | | | |
| | Nanocure TPO | | | |
| | Omnirad TPO | | | |
| | Runtecure 1108 | | | |
| | Speedcure TPO | | | |

| | | | | |
|---|---|---|---|---|
| 23 | Chivacure TPO-L<br>Hycure TPO-L<br>Lucirin TPO-L<br>Omnirad TPO-L<br>Speedcure TPO-L<br>TEPO | <br>ethyl (2,4,6-trimethylbenzoyl)phenyl-phosphinate | $C_{18}H_{21}O_3P$<br><br>316.4 | liquid<br>ca. 13 |
| 24 | Irgacure 819<br>(BAPO-2)<br>Irgacure 819DW (45% dispersion)<br>Lucirin BAPO<br>Speedcure BPO | <br>phenyl-bis-(2,4,6-trimethylbenzoyl)-phosphine oxide | $C_{26}H_{27}O_3P$<br><br>418.5 | solid<br>127–133 |
| 25 | **α-haloacetophenones and acid generators**<br>Trigonal P1 | <br>4'-(tert.butyl)-2,2,2-trichloroacetophenone | $C_{12}H_{13}Cl_3O$<br><br>279.5 | |

*(continued on next page)*

Table A.1 (continued) Type I Photoinitiators: Physico-Chemical Data

| No. | Trade name abbreviation | Structure Chemical name/Synonym | Molecular formula Mol. Wt. | Form MP/BP °C |
|-----|-------------------------|--------------------------------|----------------------------|---------------|
| 26 | Sandoray 1000 | 4'-(phenoxy)-2,2-dichloroacetophenone | $C_{14} H_{10} Cl_2 O_2$ 281.2 | solid |
| 27 | S-triazines QLCure TAZ-104 | 2-(4-methoxyphenyl)-4,6-bis-(trichloromethyl)-S-triazine | $C_{12} H_7 Cl_6 N_3 O$ 421.9 | 144–146 |
| 28 | QLCure TAZ-110 | 2,4-bis(trichloromethyl)-6-p-methoxystyryl-S-triazine | $C_{14} H_9 Cl_6 N_3 O$ 447.9 | |

| # | Name | Structure | Chemical name | Formula / MW | State | mp |
|---|------|-----------|---------------|--------------|-------|-----|
| 29 | QLCure TAZ-113 | | 2,4-bis(trichloromethyl)-6-(3,4-dimethoxy)-styryl-S-triazine | $C_{15}H_{11}Cl_6$ $N_3O_2$ 478 | | |
| 30 | QLCure TAZ-114 | | 2,4-bis(trichloromethyl)-6-(2,4-dimethoxy)-styryl-S-triazine | $C_{15}H_{11}Cl_6$ $N_3O_2$ 478 | | |
| 31 | BMPS PTBS QLCure TBPS Runtecure 1093 | | phenyl tribromomethyl sulphone | $C_7H_5Br_3$ $O_2S$ 392.9 | solid | 145–147 |

*(continued on next page)*

*Table A.1 (continued)* Type I Photoinitiators: Physico-Chemical Data

| No. | Trade name abbreviation | Structure Chemical name/Synonym | Molecular formula Mol. Wt. | Form MP/BP °C |
|-----|------------------------|--------------------------------|----------------------------|---------------|
| 32 | PAG 103 | R is propyl | $C_{16} H_{16} N_2$ $O_3 S_2$ 348.5 | solid 93–95 |
| 33 | PAG 108 | As above, R is octyl | $C_{21} H_{26} N_2$ $O_3 S_2$ 418.6 | solid 89–91 |
| 34 | PAG 121 | As above, R is p-toluenesulphonyl | $C_{20} H_{16} N_2$ $O_3 S_2$ 396.5 | solid 135–138 |
| 35 | **Specialities** Speedcure PDO | 1-phenyl-1,2-propanedione-2-(O-ethoxy-carbonyl) oxime | $C_{12} H_{13} N O_4$ 235.3 | solid 58–61 |

| 36 | Irgacure OXE-1 | 1-[4-(phenylthio)phenyl]-octane-1,2-dione-2-(O-benzoyloxime) | solid |
| 37 | Irgacure OXE-2 | 1-[9-ethyl-6-(2-methylbenzoyl)-9H-carbazole-3-yl]-ethanone-1-(O-acetyloxime) | solid |

*(continued on next page)*

*Table A.1 (continued)* Type I Photoinitiators: Physico-Chemical Data

| No. | Trade name abbreviation | Structure Chemical name/Synonym | Molecular formula Mol. Wt. | Form MP/BP °C |
|---|---|---|---|---|
| 38 | HABI Hampford Hycure BCIM Nanocure BCIM Omnirad BCIM QLCure BCIM Speedcure BCIM | 2,2'-bis(o-chlorophenyl)-4,4',5,5'-tetra-phenyl-1,2'-biimidazole | $C_{42}$ $H_{28}$ $Cl_2$ $N_4$ 663.7 | solid 196–202 |
| 39 | QLCure HABI 101 Nanocure HABI 1311 | 2,2',4-tris(2-chlorophenyl)-5-(3,4-dimethoxy phenyl)-4',5',diphenyl-1,1'-biimidazole | $C_{44}$ $H_{31}$ $Cl_3$ $N_4$ $O_2$ 754.1 | solid 90–110 |

| 40 | QLCure HABI 102 | | 2,2'-bis(o-chlorophenyl)-4,4',5,5'-tetra-(3-methoxyphenyl)biimidazole | $C_{46} H_{35} Cl_2 N_4 O_4$ 778.7 | solid |
| 41 | QLCure HABI 103 | | 2,2'-bis(o-chlorophenyl)-4,4',5,5'-tetra-(3,4,5-methoxyphenyl)biimidazole | $C_{54} H_{52} Cl_2 N_4 O_{12}$ 975.8 | solid |

*(continued on next page)*

*Table A.1 (continued)* Type I Photoinitiators: Physico-Chemical Data

| No. | Trade name abbreviation | Structure Chemical name/Synonym | Molecular formula Mol. Wt. | Form MP/BP °C |
|---|---|---|---|---|
| 42 | Irgacure 784 Titanocene | Bis (eta 2,4,5-cyclopentadien-1-yl)-bis[2,6-difluoro-3-(1Hpyrrol-1-yl)phenyl] titanium | $C_{30} H_{22} F_4$ $N_2$ Ti 534.4 | solid 160–170 |
| | Irgacure 727 | Titanocene | | |

*Table A.2* Type I Photoinitiators: Photochemical Data

| No. | Trade name abbreviation | Absorption wavelength λ max nm | | Absorption E 1% 1 cm | | Molar absorption l.M$^{-1}$ cm$^{-1}$ λ max | | Triplet energy kcal/mol |
|---|---|---|---|---|---|---|---|---|
| | | 1 | 2 | 1 | 2 | 1 | 2 | |
| | **HAPs** | | | | | | | |
| 1 | Darocur 1173 | 244 | 330 | 650 | 6 | 10,500 | 100 | 67 |
| 2 | Omnirad 102 | 256 | 325 | 650 | 6 | 13,100 | | |
| 3 | Irgacure 184 | 243 | 331 | 580 | 5 | 8,400 | 108 | 66.9 |
| 4 | Irgacure 2959 | 273 | 330 | 700 | | 15,700 | | |
| 5 | Omnirad 669 | 275 | 330 | 660 | | 14,700 | | |
| 6 | Esacure KIP 150 | 262 | 330 | | | 10,000 | 130 | 70.2 |
| 7 | Irgacure 127 | 260 | 315 | 850 | | 14,900 | | |
| 8 | Esacure ONE | 260 | 325 | 730 | | 15,000 | | |
| 9 | Irgacure PICS-1 | | | | | | | |
| | **AAAPs** | | | | | | | |
| 10 | Irgacure 907 | 230 | 303 | 350 | 750 | 8,700 | 18,600 | 61 |
| 11 | MMMP-3 | | 305 | | | | | |
| 12 | Irgacure 369 | 233 | 320 | 560 | 1030 | | 21,500 | 60 |
| 13 | BDMD-3 | | 315 | | | | | |
| 14 | Irgacure 379 | 233 | 320 | 520 | | | 20,800 | |
| 15 | Omnipol 910 | | 325 | 330 | | | 17,400 | |
| | **BKs etc.** | | | | | | | |
| 16 | Irgacure 651 | 252 | 335 | 500 | 12 | 16,510 | 266 | 66.2 |
| 17 | Uvatone 8302 | | | | | | | |
| 18 | DEAP | 245 | 335 | 570 | | 11,900 | | 73 |
| | **BEs** | | | | | | | |
| 19 | BIPE | 250 | 330 | 480 | | 12,200 | | 71.5 |
| 20 | BNBE | | | | | | | |
| 21 | BIBE | 245 | 325 | 430 | | 11,600 | 150 | |
| | **POs** | | | | | | | |
| 22 | Lucirin TPO | 275 | 379 | | 17 | | 610 | 60 |
| 23 | Lucirin TPO-L | 270 | 370 | | 7 | | 243 | |
| 24 | Irgacure 819 | 275 | 377 | 180 | 22 | | 1200 | 55.5 |

*Table A.2 (continued)* Type I Photoinitiators: Photochemical Data

| No. | Trade name abbreviation | Absorption wavelength λ max nm | | Absorption E 1% 1 cm | | Molar absorption l.M$^{-1}$ cm$^{-1}$ λ max | | Triplet energy kcal/mol |
|---|---|---|---|---|---|---|---|---|
| | | 1 | 2 | 1 | 2 | 1 | 2 | |
| | **Halo APs** | | | | | | | |
| | **Acid Gens.** | | | | | | | |
| 25 | Trigonal P1 | | | | | | | |
| 26 | Sandoray 1000 | | | | | | | |
| 27 | TAZ-104 | | 328 | | | | 27,000 | |
| 28 | TAZ-110 | | 380 | | | | | |
| 29 | TAZ-113 | | | | | | | |
| 30 | TAZ-114 | | | | | | | |
| 31 | BMPS | | | | | | | |
| 32 | PAG 103 | 255 | 405 | | | | | |
| 33 | PAG 108 | 255 | 405 | | | | | |
| 34 | PAG 121 | 255 | 405 | | | | | |
| | **Specialities** | | | | | | | |
| 35 | Speedcure PDO | 259 | | 385 | | 7,900 | | 53 |
| 36 | OXE-1 | | 325 | | | | | |
| 37 | OXE-2 | | 340 | | | | | |
| 38 | BCIM | 270 | 360 | | | | | |
| 39 | HABI 101 | | | | | | | |
| 40 | HABI 102 | | | | | | | |
| 41 | HABI 103 | | | | | | | |
| 42 | Irgacure 784 | 398 | 470 | | | | | |

*Table A.3* Type II Photoinitiators: Physico-Chemical Data

| No. | Trade name abbreviation | Structure Chemical name/Synonym | Molecular formula Mol. Wt. | Form MP/BP °C |
|---|---|---|---|---|
| | **Benzophenones** | | | |
| 50 | BP<br>Many suppliers | benzophenone | $C_{13} H_{10} O$ | solid |
| | | | 182.2 | 48-49 |
| 51 | 4-MBP<br>Ebecryl P37<br>Hycure MBP<br>Omnirad 4MBZ<br>Runtecure 1024<br>Speedcure MBP | 4-methylbenzophenone | $C_{14} H_{12} O$ | solid |
| | | | 196.2 | 56-58 |
| 52 | Kayacure MBP | 3,3'-dimethyl-4-methoxybenzophenone | $C_{16} H_{16} O_2$ | |
| | | | 240.3 | |

*(continued on next page)*

*Table A.3 (continued)* Type II Photoinitiators: Physico-Chemical Data

| No. | Trade name abbreviation | Structure Chemical name/Synonym | Molecular formula Mol. Wt. | Form MP/BP °C |
|---|---|---|---|---|
| 53 | Uvecryl P36 co-polymerisable Ebecryl P36 | $CO(OC_2H_4)_4OOC.CH=CH_2$ benzophenone-2-carboxy-(tetraethoxy)acrylate | $C_{25}H_{28}O_8$ 456.5 | liquid |
| 54 | Omnipol BP | $BP—(OC_4H_4)_nO—BP$ polyTHF-di(benzophenoneoxyacetate) | $C_{46}H_{56}O_{11}$ ca. 750 | liquid |
| 55 | Genopol BP-1 | polymeric BP | 960 | liquid |
| 56 | Speedcure 7005 | a mixture of di- and tetra-functional polymeric benzophenones | ca. 1216 | liquid |
| 57 | Goldcure 2700 | polymeric BP | | |
| 58 | Additol PBZ Hycure PBP Genocure PBZ Goldcure PBZ Omnirad 4PBZ | 4-phenylbenzophenone | $C_{19}H_{14}$ 258.3 | solid 100-102 |

| No. | Trade names | Chemical name | Formula | MW | mp / state |
|---|---|---|---|---|---|
| 59 | Runtecure 1059, Speedcure PBZ, Trigonal 12 | 4-benzoylbiphenyl | | | solid |
| 60 | Goldcure 2300 | polymeric PBZ | | | |
| | Chivacure BMS, Hycure BMDS, Omnirad BMS, Nanocure BMS, Runtecure 1030, Speedcure BMS | 4-(4'-methylphenylthio)benzophenone 4-tolylthiobenzophenone | $C_{20}H_{16}OS$ | 304.4 | 73-78 84-86 dimorphism |
| 61 | Chivacure OMB, Daitocure OB, Firstcure MOBB, Genocure MBB, Goldcure OMBB, Hycure OBM, Omnirad OMBB, Runtecure 1056, Speedcure MBB | methyl 2-benzoylbenzoate | $C_{15}H_{12}O_3$ | 240.3 | solid 48-54 |

*(continued on next page)*

*Table A.3 (continued)* Type II Photoinitiators: Physico-Chemical Data

| No. | Trade name abbreviation | Structure Chemical name/Synonym | Molecular formula Mol. Wt. | Form MP/BP °C |
|---|---|---|---|---|
| 62 | Esacure 1001M | Ph.CO.Ph.S—〈 〉—CO—C(CH₃)₂—SO₂—〈 〉—CH₃ <br><br> 1-(4-[benzoylphenylsulpho]phenyl)-2-methyl-2-(4-methylphenylsulphonyl)-propan-1-one | $C_{30} H_{26} O_4 S$ <br><br> 482.6 | solid <br><br> 110-113 |
| 63 | **Thioxanthones** <br> 2-, 4-, ITX isomers <br> Additol ITX <br> Chivacure ITX <br> Esacure ITX <br> Firstcure ITX <br> Goldcure ITX <br> Hycure ITX <br> Nanocure ITX <br> Runtecure 1105 <br> SarCure SR 1124 <br> Speedcure ITX | <br> 2- and 4-isopropylthioxanthone | $C_{16} H_{14} O S$ <br><br> 254.3 | solid <br><br> 57-74 |

| 64 | 2-ITX<br>Chivacure 2-ITX<br>Genocure ITX<br>Omnirad ITX<br>Speedcure 2-ITX | 2-isopropylthioxanthone | $C_{16} H_{14} O S$<br>254.3 | solid<br>74-76 |
| 65 | Chivacure DETX<br>Genocure DETX<br>Kayacure DETX<br>Nanocure DETX<br>Omnirad DETX<br>Runtecure DETX<br>Speedcure DETX | 2,4-diethylthioxanthone | $C_{17} H_{16} O S$<br>268.4 | solid<br>66-70 |
| 66 | Omnirad CTX<br>Kayacure CTX<br>Speedcure CTX | 2-chlorothioxanthone | $C_{13} H_7 Cl O S$<br>246.7 | solid<br>152-153 |
| 67 | Kayacure RTX | 2,4-dimethylthioxanthone | $C_{15} H_{12} O S$ | solid |

*(continued on next page)*

*Table A.3 (continued)* Type II Photoinitiators: Physico-Chemical Data

| No. | Trade name abbreviation | Structure Chemical name/Synonym | Molecular formula Mol. Wt. | Form MP/BP °C |
|---|---|---|---|---|
| 68 | Kayacure DITX | 2,4-diisopropylthioxanthone | $C_{19} H_{20} O S$ | solid |
| 69 | Speedcure CPTX | 1-chloro-4-propoxythioxanthone | $C_{16} H_{18} Cl O_2 S$ 304.8 | solid 99–103 |
| 70 | Speedcure 7010 | polymeric CPTX (dendrimer) | c 1800 | solid 73 |
| 71 | Omnipol TX polymeric | TX—$(OC_4H_8)_n$O—TX polyTHF-di(thioxanthone-2-oxyacetate) (linear polymer) | $C_{46} H_{50} O_{12} S_2$ ca. 800 | resin |
| 72 | Genopol TX-1 | polymeric TX (branched) | 820 | |

| | | | | |
|---|---|---|---|---|
| 73 | **Miscellaneous**<br>MBF, MPG<br>Additol MBF<br>Chivacure 200<br>Darocur MBF<br>Genocure MBF<br>Hycure MBF<br>Omnirad MBF<br>Runtecure 1055<br>Speedcure MBF | —COOCOCH$_3$<br><br>methyl benzoylformate<br>methyl phenylglyoxylate | C$_9$H$_8$O$_3$<br><br>164.2 | liquid<br>MP 17<br>BP 242-250 |
| 74 | Irgacure 754 | $\overset{O}{\overset{\|}{C}}.\overset{O}{\overset{\|}{C}}.O-[CH_2CH_2O]_n-R$<br>polyethylene glycol di(benzoylformate)<br>(mixed esters)         R = PhCOCO- | ca. 600 | liquid |
| 75 | CQ<br>Hampford | dl-camphorquinine<br>2,3-bornanedione | C$_{10}$H$_{14}$O$_2$<br><br>166.2 | solid<br>198-200 |

(continued on next page)

Table A.3 (continued) Type II Photoinitiators: Physico-Chemical Data

| No. | Trade name abbreviation | Structure Chemical name/Synonym | Molecular formula Mol. Wt. | Form MP/BP °C |
|---|---|---|---|---|
| 76 | benzil | benzil | $C_{14} H_{10} O_2$ <br> 210.2 | solid <br> 94-95 |
| 77 | Hycure EAQ <br> Hampford <br> Speedcure EAQ | 2-ethylanthraquinone <br> 2-ethyl-9,10-anthracene dione | $C_{16} H_{12} O_2$ <br> 236.3 | solid <br> 108-111 |

| | | | | |
|---|---|---|---|---|
| 78 | Omnipol SZ (Type II sensitiser) | polyethyleneglycol-bis-[(4-acetylphenyl)-piperazine propionate] | $C_{36} H_{50} N_4 O_8$ 716 | liquid |
| 79 | Fluorenone (sensitiser) | 9-fluorenone | $C_{13} H_8 O$ 180.2 | solid 82-85 |

*Table A.4* Type II Photoinitiators: Photochemical Data

| No. | Trade name abbreviation | Absorption wavelength $\lambda$ max | | Absorption E 1%, 1 cm | | Molar absorption $1.M^{-1}\ cm^{-1}$ $\lambda$ max | | Triplet energy kcal/mol |
|---|---|---|---|---|---|---|---|---|
| | | 1 | 2 | 1 | 2 | 1 | 2 | |
| | **BPs** | | | | | | | |
| 50 | BP | 248 | 338 | 550 | 8 | 12,000 | 130 | 69.1 |
| 51 | MBP | 245 | 330 | 534 | | | | 69 |
| 52 | Kayacure MBP | 295 | | 720 | | | | |
| 53 | Uvecryl P36 | 255 | 320 | 370 | | | | |
| 54 | Omnipol BP | 280 | 325 | 360 | | | | |
| 55 | Genopol BP-1 | 245 | 325 | | | | | |
| 56 | Speedcure 7005 | 245 | 282 | 312 | 53 | | | |
| 57 | Goldcure 2700 | | | | | | | |
| 58 | Speedcure PBZ | 283 | | 960 | | 20,900 | | 60.7 |
| 59 | Goldcure 2300 | | | | | | | |
| 60 | Speedcure BMS | 250 | 312 | 439 | 576 | | 17,500 | 61 |
| 61 | Genocure MBB | 253 | 282 | 440 | 110 | 15,200 | | |
| 62 | Esacure 1001M | 248 | 318 | 480 | 360 | 24,500 | 18,300 | 66 |
| | **TXs** | | | | | | | |
| 63 | ITX 2/4 isomers | 258 | 383 | 1120 | 168 | 49,700 | 5,900 | 61.4 |
| 64 | 2-ITX | 255 | 384 | | 170 | | 5,900 | 61.4 |
| 65 | DETX | 261 | 384 | 970 | 172 | | 5,800 | |
| 66 | CTX | 260 | 385 | 830 | 159 | | 5,700 | 63.3 |
| 67 | RTX | | | | | | | |
| 68 | DITX | | | | | | | |
| 69 | Speedcure CPTX | 312 | 387 | 277 | 175 | | 6,100 | 60.3 |
| 70 | Speedcure 7010 | 313 | 385 | 193 | 107 | | | |
| 71 | Omnipol TX | 225 | 390 | 180 | | | | |
| 72 | Genopol TX-1 | 315 | 375 | | | | | |
| | **Miscellaneous** | | | | | | | |
| 73 | Genocure MBF | 254 | 325 | 478 | 3 | 11,930 | 70 | 65.7 |
| 74 | Irgacure 754 | 254 | 325 | 650 | 3 | 12,000 | | |
| 75 | CQ | 468 | | | | 50 | | 50-57 |
| 76 | benzil | 260 | | 1000 | | 22,000 | | |
| 77 | EAQ | 257 | 334 | 1613 | 183 | | 3,500 | |
| 78 | Omnipol SZ | | 330 | | 550 | | 19,700 | |
| 79 | fluorenone | | 390 | | | | | |

*Table A.5* Eutectic Blends and Synergistic Mixtures

| No. | Trade name abbreviation | Mixture | | | Ratio % | Form |
| --- | --- | --- | --- | --- | --- | --- |
| | | 1 | 2 | 3 | | |
| 80 | Chivacure LBP Esacure TZM Genocure LBP Omnirad BEM PI 81 Runtecure 1041 Speedcure BEM | BP | MBP | | 50/50 | liquid |
| 81 | Esacure HB Genocure LBC Irgacure 500 Omnirad 500 PI 500 Runtecure 1500 | BP | 184 | | 50/50 | liquid |
| 82 | Irgacure 4665 Speedcure 210 | BP | 1173 | | 50/50 | liquid |
| 83 | Irgacure 1000 Omnirad 1000 PI-1000 Runtecure 1100 | 1173 | 184 | | 80/20 | liquid |
| 84 | Darocur 4265 Genocure LTM IGM 4265 Runtecure 1265 Speedcure 220 | 1173 | TPO | | 50/50 | liquid |
| 85 | Irgacure 149 | 1173 | BAPO-1 | | 95/5 | liquid |
| 86 | Irgacure 1700 | 1173 | BAPO-1 | | 75/25 | liquid |
| 87 | Irgacure 1870 | HAP | BAPO | | | |
| 88 | Irgacure 2022 | 1173 | 819 | | 80/20 | liquid |
| 89 | Darocur 4263 | 1173 | TPO | | 85/15 | liquid |
| 90 | Darocur 4043 | 1173 | ITX | DMB | | liquid |
| 91 | Esacure KTO 46 | 1173 | TPO | TZT | | liquid |
| 92 | Esacure DP 250 | KTO 46 | water | emulsion | | |
| 93 | Irgacure 1800 | 184 | BAPO-1 | | 75/25 | solid |
| 94 | Irgacure 1850 | 184 | BAPO-1 | | 50/50 | |
| 95 | Irgacure 2100 | 819 mix | | | | liquid |

*(continued on next page)*

*Table A.5 (continued)* Eutectic Blends and Synergistic Mixtures

| No. | Trade name abbreviation | Mixture | | | Ratio % | Form |
|-----|------------------------|---------|---|---|---------|------|
| | | 1 | 2 | 3 | | |
| 96 | SarCure SR 1135 | TPO | SR 1130 | SR 1137 | | liquid |
| 97 | SarCure SR 1126 | SR 1135 | water | emulsion | | liquid |
| 98 | Irgacure 1300 | BDK | 369 | | 70/30 | solid |
| 99 | Esacure TZT SarCure SR 1137 | Trimethyl BP | MBP | eutectic | 80/20 | liquid |
| 100 | Esacure KT37 SarCure SR 1133 | KIP 150 | TZT | | 30/70 | liquid |
| 101 | Esacure KIP 100F SarCure SR 1129 | KIP 150 | 1173 | | 70/30 | liquid |
| 102 | Esacure KIP 75LT | KIP 150 | TPGDA | | 75/25 | liquid |
| 103 | Esacure KT 55 | KIP 150 | TZT | | 50/50 | liquid |
| 104 | Esacure KIP IT | KIP 150 | GPTA | | 65/35 | liquid |
| 105 | SarCure SR 1132 | KIP 150 | TMPTA | | | liquid |
| 106 | Esacure KIP EM Sarcure SR 1131 | KIP 150 | water | emulsion | | liquid |
| 107 | Irgacure 819DW | 819 | water | dispersion | 45% 819 | liquid |

*Note:*  Many companies offer blends for particular applications.

*Table A.6* Hydrogen Donors, Tertiary Amines, Thiols, etc.

| No. | Trade name abbreviation | Structure Chemical name | Molecular formula Mol. Wt. | Form MP/BP °C | UV abs. l max nm |
|---|---|---|---|---|---|
| | **Tertiary amines** | | | | |
| 110 | TEA | $N(CH_2CH_3)_3$ triethylamine | $C_6H_{15}N$ 101.2 | liquid 89 | |
| 111 | DMEA | $(CH_3)_2NCH_2CH_2OH$ N,N-dimethylethanolamine 2-dimethylaminoethanol | $C_4H_{11}NO$ 89.1 | liquid 133–134 | |
| 112 | MDEA Genocure | $CH_3N(CH_2CH_2OH)_2$ N-methyldiethanolamine | $C_5H_{13}NO_2$ 119.2 | liquid 246–248 | 220 |
| 113 | TEOA | $N(CH_2CH_2OH)_3$ triethanolamine | $C_6H_{15}NO_3$ 149.2 | liquid 190–193 5 mm | |

(continued on next page)

*Table A.6 (continued)* Hydrogen Donors, Tertiary Amines, Thiols, etc.

| No. | Trade name abbreviation | Structure Chemical name | Molecular formula Mol. Wt. | Form MP/BP °C | UV abs. l max nm |
|-----|------------------------|------------------------|---------------------------|---------------|-------------------|
| 114 | DMB IGM Speedcure | —COOCH$_2$CH$_2$N(CH$_3$)$_2$<br><br>2-(dimethylamino)ethyl benzoate | C$_{11}$H$_{15}$NO$_2$<br><br>193.2 | liquid<br><br>107–108<br>2 mm | 270 |
| 115 | **Aminobenzoates**<br>EDB or EPD<br>Additol EPD<br>Chivacure<br>Esacure EDB<br>Genocure EPD<br>Hycure EPD<br>Kayacure EPA<br>Nanocure EPD<br>Omnirad EDB<br>Runtecure 1101<br>SarCure SR 1125<br>Speedcure EDB | (CH$_3$)$_2$N—⟨⟩—COOC$_2$H$_5$<br><br>ethyl 4-(dimethylamino)benzoate | C$_{11}$H$_{15}$NO$_2$<br><br>193.2 | solid<br><br>62–64 | 315 |

| 116 | EHA<br>Additol EHA<br>Chivacure OPD<br>Esacure EHA<br>Escalol 507<br>Firstcure ODAB<br>Genocure EHA<br>Hycure EHA<br>Nanocure EHA<br>Runtecure 1098<br>Speedcure EHA | (CH$_3$)$_2$N—⬡—COOCH$_2$CHC$_4$H$_9$ \| CH$_2$CH$_3$<br><br>2-ethylhexyl-4-dimethylamino-benzoate | C$_{17}$H$_{27}$NO$_2$<br>277.4 | liquid<br>185–195<br>5 mm | 310 |
| --- | --- | --- | --- | --- | --- |
| 117 | IGM IADB<br>Kayacure DMBI | (CH$_3$)$_2$N—⬡—COOC$_5$H$_{11}$<br><br>isoamyl 4-(dimethylamino)-benzoate | C$_{14}$H$_{21}$NO$_2$<br>235.3 | liquid<br>150–155<br>2.5 mm | |
| 118 | BEDB<br>Speedcure | (CH$_3$)$_2$N—⬡—COOC$_2$H$_4$OC$_4$H$_9$<br><br>2-butoxyethyl 4-(dimethylamino)-benzoate | C$_{15}$H$_{23}$NO$_3$<br>265.3 | liquid<br>178–181<br>1.5 mm | 310 |

*(continued on next page)*

*Table A.6 (continued)* Hydrogen Donors, Tertiary Amines, Thiols, etc.

| No. | Trade name abbreviation | Structure Chemical name | Molecular formula Mol. Wt. | Form MP/BP °C | UV abs. 1 max nm |
|---|---|---|---|---|---|
| 119 | Speedure 7040 | di- and tetra-functional polymeric aminobenzoate | c 1040 | liquid | 311 |
| 120 | Omnipol ASA | polymeric aminobenzoate | | solid | |
| 121 | Omnipol ASE | polymeric aminobenzoate | | | |
| 122 | Esacure A198 | polymeric aminobenzoate | | | |
| 123 | Genopol AB-1 | polymeric aminobenzoate | | | 315 |
| 124 | Genopol RCX02 | polymeric aminobenzoate | | | 310 |
| 125 | **Miscellaneous** MK Michler's ketone | $(CH_3)_2N$ ⬡ -CO- ⬡ $-N(CH_3)_2$ 4,4′-bis(dimethylamino)-benzophenone | $C_{17}H_{20}N_2O$ 268.4 | solid 174–176 | 370 |

| 126 | EMK<br>Chivacure<br>IGM<br>Hycure<br>Nanocure<br>Omnirad<br>Speedcure<br>Firstcure DEAB | (C$_2$H$_5$)$_2$N—⟨ ⟩—CO—⟨ ⟩—N(C$_2$H$_5$)$_2$<br>4,4'-bis(diethylamino)-benzophenone<br>ethyl Michlers ketone | C$_{21}$H$_{28}$N$_2$O<br>324.5 | solid<br>94–96 | 376 |
|---|---|---|---|---|---|
| 127 | P101, P104<br>Many sources | acrylated amines<br>oligoamines | various | liquid | |
| 128 | NPG | —NHCH$_2$COOH<br>N-phenylglycine | C$_8$H$_9$NO$_2$<br>151.2 | solid<br>126–127 | |

*(continued on next page)*

*Table A.6 (continued)*  Hydrogen Donors, Tertiary Amines, Thiols, etc.

| No. | Trade name abbreviation | Structure Chemical name | Molecular formula Mol. Wt. | Form MP/BP °C | UV abs. l max nm |
|---|---|---|---|---|---|
| 129 | LCV IGM Speedcure | $(CH_3)_2N$ — $N(CH_3)_2$ — $(NCH_3)_2$ — CH 4,4′,4″-tris(dimethylamino)-triphenylmethane | 373.5 | solid 170–177 | 208 |
| 130 | **Thiols** thiol | $HSCH_2CH_2COO(CH_2)_3CH_3$ butyl 3-mercaptopropionate | 162.2 | liquid 101/12 mm | |

| No. | Type | Structure | Name | Formula | MW | State | bp/mp |
|---|---|---|---|---|---|---|---|
| 131 | Trithiol | | Trimethylolpropane-tris(2-mercaptoacetate) | $C_{12}H_{20}O_6S_3$ | 356.5 | liquid | 245/1 mm |
| 132 | Trithiol | | trimethylolpropane-tris(2-mercaptopropionate) | $C_{15}H_{26}O_6S_3$ | 398.6 | liquid | 220/0.3 mm |
| 133 | thiol | | 2-mercaptobenzothiazole<br>2-benzothiazolethiol | $C_7H_5NS_2$ | 167.2 | solid | 177–181 |

*(continued on next page)*

*Table A.6 (continued)* Hydrogen Donors, Tertiary Amines, Thiols, etc.

| No. | Trade name abbreviation | Structure Chemical name | Molecular formula Mol. Wt. | Form MP/BP °C | UV abs. l max nm |
|-----|-------------------------|-------------------------|----------------------------|----------------|------------------|
| 134 | thiol | 2-mercaptobenzimidazole 2-benzimidazolethiol | $C_7 H_6 N_2 S$ 150.2 | solid 301–305 | |

*Table A.7* Cationic Photoinitiators: Structures

| No. | Trade Name abbreviation | Structure Chemical name/Synonym | Anion | Benzene release |
|-----|------------------------|--------------------------------|-------|-----------------|
| 140 | **Triarylsulphonium salts**<br>Cyracure UVI-6976<br>QL Cure 201<br>Speedcure 976<br><br>(mixed salts) | <br>diphenyl-(4-phenylthio)-phenyl-sulphonium hexafluoroantimonate<br>(Structure 2) | $SbF_6^-$ | Yes |

*(continued on next page)*

*Table A.7 (continued)* Cationic Photoinitiators: Structures

| No. | Trade Name abbreviation | Structure Chemical name/Synonym | Anion | Benzene release |
|-----|------------------------|--------------------------------|-------|-----------------|
| | | thio-di-1,4-phenylene-bis(diphenyl-sulphonium) hexafluoroantimonate (Structure 3) | | |
| 141 | Cyracure UVI-6992 Esacure 1064 Omnicat 432 QL Cure 202 Speedcure 992 | Mixed salts as above | $PF_6^-$ | Yes |

| | | | | |
|---|---|---|---|---|
| 142 | Degacure K185<br>Sarcat K185<br>SP-55 | thio-di-1,4-phenylene-bis(diphenyl-sulphonium) hexafluorophosphate | $PF_6^-$ | Yes |
| 143 | QL Cure 211<br>(mixed salts) | di-(4-methylphenyl)4-(4-methylphenyl-thio)phenylsulphonium hexafluoroantimonate (Structure 2) | $SbF_6^-$ | No |

*(continued on next page)*

*Table A.7 (continued)* Cationic Photoinitiators: Structures

| No. | Trade Name abbreviation | Structure Chemical name/Synonym | Anion | Benzene release |
|---|---|---|---|---|
| | | thio-di-1,4-phenylene-bis{di-(4-methyl-phenyl)sulphonium} hexafluoroantimonate (Structure 3) | | |
| 144 | QL Cure 212 | mixed salts as QL211 | $PF_6^-$ | No |

| | | | | |
|---|---|---|---|---|
| 145 | SP-150 | thio-di-1,4-phenylene-bis{di-(4-hydroxy-ethoxyphenyl)sulphonium} hexafluorophosphate | $PF_6^-$ | No |
| 146 | SP-170 | As SP-150, SbF6- salt. | $SbF_6^-$ | No |
| 147 | Omnicat 550 | 10-biphenyl-4-yl-2-isopropyl-9-oxo-9H-thioxanthen-10-ium hexafluorophosphate | $PF_6^-$ | No |

*(continued on next page)*

Table A.7 (continued) Cationic Photoinitiators: Structures

| No. | Trade Name abbreviation | Structure Chemical name/Synonym | Anion | Benzene release |
|---|---|---|---|---|
| 148 | Omnicat 555 | 40% solution of 550 in monomer | $PF_6^-$ | No |
| 149 | Omnicat 650 | Polymeric version of 550 | | |
| 150 | Esacure 1187 | 9-(4-hydroxyethoxyphenyl)thianthrenium hexafluorophosphate | | |
| 151 | **Diaryliodonium salts** Hycure 810 | diphenyliodonium hexafluorophosphate | $PF_6^-$ | Yes |

| | | | | |
|---|---|---|---|---|
| 152 | Uvacure 1600 | (4-octyloxyphenyl)phenyliodonium hexafluoroantimonate | $SbF_6^-$ | Yes |
| 153 | Sarcat CD-1012 | {4-[(2-hydroxy-tetradecyl)oxy]phenyl]-phenyliodonium hexafluoroantimonate | $SbF_6^-$ | Yes |
| 154 | Omnicat 440 Hycure 820 | bis-(4-methylphenyl)iodonium hexafluorophosphate | $PF_6^-$ | No |
| | Omnicat 445 | 50% solution of 440 in 3-ethyl-3-hydroxy-methyloxetane | | |

*(continued on next page)*

*Table A.7 (continued)* Cationic Photoinitiators: Structures

| No. | Trade Name abbreviation | Structure Chemical name/Synonym | Anion | Benzene release |
|---|---|---|---|---|
| 155 | Irgacure 250 | 4-isobutylphenyl-4′-methylphenyliodonium hexafluorophosphate | PF6 - | No |
| 156 | UV 9310 | bis-(4-dodecylphenyl)iodonium hexafluoroantimonate | $SbF_6^-$ | No |
| 157 | Rhodorsil 2074 | tolylcumyliodonium tetrakis-(pentafluorophenyl) borate | Borate | No |

| # | Name | Structure | | |
|---|------|-----------|---|---|
| 158 | Rhodorsil 2076 | 4-isopropylphenyl-4′-methylphenyl iodonium hexafluorophosphate | $PF_6^-$ | No |
| 159 | **Ferrocenium salts**<br>Irgacure 261 | cyclopentadienylcumene iron hexafluorophosphate | $PF6_6^-$ | No |
| 160 | **Sensitizers**<br>Anthracure UVS-1101<br>DEA | 9,10-diethoxyanthracene | | |

*(continued on next page)*

*Table A.7 (continued)* Cationic Photoinitiators: Structures

| No. | Trade Name abbreviation | Structure Chemical name/Synonym | Anion | Benzene release |
|-----|-------------------------|----------------------------------|-------|-----------------|
| 161 | Anthracure UVS-1331 DBA | 9,10-dibutoxyanthracene | | |
| 162 | EDMA (laboratory chemical) | 2-ethyl-9,10-dimethoxyanthracene | | |
| 69  | CPTX | | | |

*Table A.8* Cationic Photoinitiators. UV Absorption Data

| No. | Trade name | Molecular formula Mol. Wt. | Form MP/BP °C | Absorption wavelength | |
|---|---|---|---|---|---|
| | | | | λ max nm 1 | λ max nm 2 |
| | **Triphenylsulphonium salts** | | | | |
| 140 | Cyracure UVI-6976 | (2) $C_{24} F_6 H_{19} S_2 Sb$ 607.4 | 50% soln in PC | 225 | 300 |
| | | (3) $C_{36} F_6 H_{28} S_3 Sb$ 792.6 | | | |
| 141 | Cyracure UVI-6992 | (2) $C_{24} F_6 H_{19} S_2 P$ 516.6 | 50% soln in PC | 225 | 295 |
| | | (3) $C_{36} F_6 H_{28} S_3 P$ 702.6 | | | |
| 142 | Degacure K185 | $C_{36} F_6 H_{28} S_3 P$ 702.6 | 33% soln in PC | 225 | 295 |
| 143 | QL Cure 211 | (2) $C_{27} F_6 H_{25} S_2 Sb$ 649.4 | | 225 | 310 |
| | | (3) $C_{40} F_6 H_{36} S_3 Sb$ 848.7 | | | |

*(continued on next page)*

Table A.8 (continued) Cationic Photoinitiators. UV Absorption Data

| No. | Trade name | Molecular formula Mol. Wt. | Form MP/BP °C | Absorption wavelength | |
|---|---|---|---|---|---|
| | | | | $\lambda$ max nm 1 | $\lambda$ max nm 2 |
| 144 | QL Cure 212 | (2) $C_{27} F_6 H_{25} S_2 P$ 558.6 (3) $C_{40} F_6 H_{36} S_3 P$ 757.9 | | 225 | 310 |
| 145 | SP-150 | $C_{42} F_6 H_{44} O_8 S_3 P$ 918.1 | 27% soln in PC | 230 | 280 |
| 146 | SP-170 | $C_{42} F_6 H_{44} O_8 S_3 Sb$ 1008.9 | 27% soln in PC | 230 | 280 |
| 147 | Omnicat 550 | $C_{28} F_6 H_{22} O P S$ 541.6 | | | 285 |
| 148 | Omnicat 555 | Solution of Omnicat 550 in monomer | 40% soln | | 285 |
| 149 | Omnicat 650 | | | | 250 |
| 150 | Esacure 1187 | $C_{14} F_6 H_{13} O_2 P S_2$ 422.4 | 75% soln in PC | 250 | 310 |

| | | | | |
|---|---|---|---|---|
| | **Diphenyliodonium salts** | | | |
| 151 | Hycure 810 | $C_{12} F_6 H_{10}$ I P<br>426.1 | solid | 227 |
| 152 | Uvacure 1600 | $C_{20} F_6 H_{26}$ I Sb<br>629.1 | solid<br>57-58 | 240 |
| 153 | Sarcat CD-1012 | $C_{26} F_6 H_{38}$ I P<br>713.3 | | 250 |
| 154 | Omnicat 440 | $C_{14} H_{14} F_6$ I P<br>454.1 | solid<br>175-180 | 240 |
| | Omnicat 445 | Solution of Omnicat<br>440 in oxetane | 50% soln | 240 |
| 155 | Irgacure 250 | $C_{17} F_6 H_{20}$ I P<br>496.3 | 75% in PC | 240 |
| 156 | UV 9310 | $C_{36} H_{58}$ I Sb<br>739.6 | | |

*(continued on next page)*

*Table A.8 (continued)* Cationic Photoinitiators. UV Absorption Data

| No. | Trade name | Molecular formula Mol. Wt. | Form MP/BP °C | Absorption wavelength | |
|---|---|---|---|---|---|
| | | | | $\lambda$ max nm 1 | $\lambda$ max nm 2 |
| 157 | Rhodorsil 2074 | $C_{40} H_{18} B F_{20} I$ 1016.3 | solid 120-133 or 18% soln in ipa 50% in PC | 240 | 270 |
| 158 | Rhodorsil 2076 | | | 240 | |
| 159 | **Ferrocenium salts** Irgacure 261 | $C_{14} F_6 H_{18} Fe P$ 387.3 | 75% soln | 240 | |
| 160 | **Sensitizers** Anthracure UVS-1101 DEA | $C_{18} H_{18} O_2$ 266.3 | solid 141 | 380 | 402 |
| 161 | Anthracure UVS-1331 DBA | $C_{22} H_{26} O_2$ 322.4 | | 380 | 403 |
| 162 | EDMA | $C_{18} H_{18} O_2$ 266.3 | solid 117–119 | | 401 |

*Table A.9* Solubilities. % Wt/vol in Solvents and Monomers

| No. | Trade name abbreviation | Acetone | Ethanol | Toluene | HDDA | TPGDA | TMPTA |
|-----|------------------------|---------|---------|---------|------|-------|-------|
| 1 | Darocur 1173 | >50 | | >50 | >50 | >50 | >50 |
| 2 | Omnirad 102 | >50 | | >50 | >50 | >50 | >50 |
| 3 | Irgacure 184 | >50 | >50 | >50 | >50 | >50 | >50 |
| 4 | Irgacure 2959 | 20 | | | 10 | 20 | 5 |
| 5 | Omnirad 669 | | | | poor | | |
| 6 | Esacure KIP 150 | | | | poor | | |
| 7 | Irgacure 127 | | | | | 15 | 5 |
| 8 | Esacure ONE | | | | | | |
| 9 | Irgacure PICS-1 | | | | | | |
| 10 | Irgacure 907 | >50 | | >50 | 30 | 22 | 20 |
| 11 | MMMP-3 | | | | misc. | misc. | misc. |
| 12 | Irgacure 369 | 17 | | 27 | 5 | 6 | 5 |
| 13 | BDMD-3 | | | | misc. | misc. | misc. |
| 14 | Irgacure 379 | >50 | | | 30 | 24 | 20 |
| 15 | Omnipol 910 | | | | | | |
| 16 | Irgacure 651 | >50 | >50 | >50 | 30 | 30 | 25 |
| 18 | DEAP | | | | | | |
| 21 | BE | | | | | | |
| 22 | Lucirin TPO | 50 | | | 22 | 16 | 14 |
| 23 | Lucirin TPO-L | | | | >50 | >50 | >50 |
| 24 | Irgacure 819 | 14 | | 20 | 10 | 5 | 5 |
| 27 | TAZ 104 | | | | | | |
| 28 | Taz 110 | | | | | | |
| 35 | Speedcure PDO | 100 | 10 | 80 | 40 | | 14 |
| 38 | BCIM | | | | | | |
| 42 | Irgacure 784 | 30 | | | 10 | 5 | 5 |
| 50 | BP | >50 | >50 | >50 | 40 | 37 | 30 |
| 51 | Speedcure MBP | | | | | | |
| 52 | Kayacure MBP | | | | | | |
| 53 | Uvecryl P36 | | | | | | |
| 54 | Omnipol BP | | | | | | |
| 55 | Genopol BP-1 | | | | | | |
| 56 | Speedcure 7005 | | | | | | |
| 58 | Speedcure PBZ | | 1 | 14 | 5 | 5 | 3 |
| 59 | Goldcure 2300 | | | | | | |
| 60 | Speedcure BMS | 35 | 2 | 40 | 14 | | 3 |

*(continued on next page)*

*Table A.9 (continued)* Solubilities. % Wt/vol in Solvents and Monomers

| No. | Trade name abbreviation | Acetone | Ethanol | Toluene | HDDA | TPGDA | TMPTA |
|---|---|---|---|---|---|---|---|
| 61 | Genocure MBB | | | | | | |
| 62 | Esacure 1001M | | | | | | |
| 63 | 2,4-ITX | 30 | 4 | 30 | 30 | 15 | 10 |
| 64 | 2-ITX | | | | | | |
| 65 | DETX | 40 | | | 35 | 20 | 15 |
| 66 | CTX | 2 | 1 | 4 | 4 | | 2 |
| 69 | Speedcure CPTX | 5 | 1 | 10 | 6 | 5 | 3 |
| 70 | Speedcure 7010 | | | | | | |
| 71 | Omnipol TX | | | | | | |
| 72 | Genopol TX-1 | | | | | | |
| 73 | Genocure MBF | >50 | | | >50 | >50 | >50 |
| 74 | Irgacure 754 | >50 | | | >50 | >50 | >50 |
| 75 | CQ | | sol | | | | |
| 76 | Benzil | | | | | | |
| 77 | EAQ | | | | | | |
| 78 | Omnipol SZ | | | | | | |
| 114 | DMB | | >50 | | >50 | | >50 |
| 115 | EDB | >50 | | | 30 | | 20 |
| 116 | EHA | | | | | | |
| 126 | EMK | | | | | | |
| 128 | NPG | sol | sol | sol | | | |
| 129 | LCV | | | | | | |

*Table A.10* Trademarks and Suppliers

| Trademark | Company |
|---|---|
| Additol | Cytec Industries Inc.<br>5 Garret Mountain Plaza, West Paterson, NJ, 07424, USA<br>Tel. 937 357 3100<br>www.cytec.com |
| Anthracure | Kawasaki Kasei Chemicals Ltd.<br>Kawasakiekimea Tower River Bldg. 12-1, Ekimaehoncho,<br>   Kawasaki-ku, Kawasaki City, Kanagawa 210-0007 Japan<br>Tel. 81 44 246 7454<br>www.quinone.com |
| Chivacure | Chitec Technology Co. Ltd.<br>7F, 58, Lane 148, Li De Street, Chung Ho, Taipei Hsien 235, Taiwan |
| Cyracure | Dow |
| Daitocure | Siber Hegner |
| Darocur | Ciba (formally Merck)<br>See Irgacure |
| Degacure | Degussa AG, now marketed by Sartomer |
| Esacure | Lamberti Spa<br>Via Marsala 38/D, 21013 Gallerate Va, Italy<br>Tel. +39 0331 715780<br>www.esacure.com |
| Firstcure | Albermarle Corporation<br>451 Florida St., Baton Rouge, LA, 7081-1765, USA<br>Tel. 225 388 7402<br>www.albermarle.com |
| Genocure<br>   Genopol | Rahn AG<br>Rahn USA Corp.<br>1005 N. Commons Drive, Aurora, IL 60504, USA<br>Tel. 630 851 4220<br>www.rahn-group.com |
| Genorad | Dorflistrasse 120 CH-8050 Zurich<br>www.rahn-group.com<br>Tel. +41(0) 44 315 4200 |
| Hicure | Kawaguchi |

*(continued on next page)*

*Table A.10 (continued)* Trademarks and Suppliers

| Trademark | Company |
| --- | --- |
| Hycure | ChemFine<br>International Co. Ltd.<br>99 Qingyang Rd, Wuxi, China<br>Tel. 0086 510 827 53588<br>www.chemfineinternational.com |
| Irgacure | Ciba Speciality Chemicals Inc.<br>PO Box CH-4002, Basel, Switzerland<br>www.cibasc.com |
| Kayacure | Nippon Kayaku<br>www.nipponkayaku.co.jp |
| Lucirin | BASF Aktiengesellschaft<br>67056 Ludwigshafen, Germany<br>www.basf.com |
| Micure | Miwan |
| Nanocure | Daelim Chemical Co. Ltd.<br>660-903 233-7 Sangpyeong-dong, Jinju, Kyeongsangnam-do,<br>  Korea<br>Tel. +82 2589 0868<br>www.dichem.co.kr |
| Omnirad<br>  Omnicat<br>  Omnipol | IGM Resins<br>Vredesplein 28, PO Box 265, 5140 AG Waalwijk, the Netherlands<br>Tel. +31 (0) 416 316 657<br>www.igmresins.com |
| Quadracure | Chitec Technology Co. Ltd. |
| Quantacure | Ward Blenkinsop (later Great Lakes Fine Chemicals)<br>Now marketed under Speedcure |
| QLCure | Quang Li Chemical Co. Ltd.<br>Quianjia Industry Park, Yaoguan Town<br>Quianjia Industry Park, Yaoguan Town, Changzhou, Jiangsu,<br>  China<br>Tel. 86 0519 838 1281<br>www.qlsyn.com |
| Rhodorsil | Rhodia. S.A.<br>110 Esplanade Charles de Gaulle, La Defense, Paris, France<br>Tel. +33 (1) 5356 6464<br>www.rhodia.com |

*Table A.10 (continued)* Trademarks and Suppliers

| Trademark | Company |
|---|---|
| Runtecure | Runtec Chemical Co. Ltd.<br>166 Heng Street, Jintai, Jiangsu, 21320 China<br>Tel. +86 519 329 8808<br>www.runtec-chem.com |
| Sandoray | Sandoz |
| SarCure<br>Sarcat | Sartomer Company Inc.<br>502 Thomas Jones Way, Exton, PA 19341, USA<br>Tel. 610 363 4100<br>www.sartomer.com |
| Speedcure | Lambson Ltd. Lamro House<br>603 Thorp Arch Estate, Wetherby, LS23 7FS, UK<br>Tel. 0044(0) 1937 840170<br>www.lambson.com |
| Trigonal | Akzo |
| Uvacure | Surface Specialities (UCB)<br>1950 Lake Park Dr., Smyrna, GA 30080, USA<br>Tel. 800 433 2873<br>www.surfacespecialities.com |
| Uvatone | Upjohn |
| UVI | Dow Chemical Co.<br>Midland, MI 48674, USA<br>Tel. 800 447 4369 |
| Vicure | Stauffer |
| Also | New Sun Corporation Ltd.<br>Po Box 19-1. Liao Jia Du, Wu Jia Ling, Changsha, Hunan 410003 China<br>Tel. +86 731 425 2605<br>www.newsunchem.com<br><br>Hampford Research Inc.<br>54 Veterans Blvd, Po Box 1073, Stratford, CT 06615, USA<br>Tel. 203 375 1137<br><br>Hunan Xinyu Co. Ltd.<br>183 Yuandayi Road, Changsha, China<br>Tel. +86 731 442 1482 |

*(continued on next page)*

*Table A.10 (continued)* Trademarks and Suppliers

| Trademark | Company |
|-----------|---------|
|           | Aceto Corporation<br>One Hollows Lane, Lake Success, NY, 11042-1215, USA<br>Tel. 516 627 6000<br>www.aceto.com |
|           | Biddle Sawyer Corp.<br>21 Penn Plaza, 360 West 31st Street, NY, 10001-2727, USA<br>Tel. 212 736 1580<br>www.biddlesawyer.com |

*Table A.11* Registrations

| No. | Trade name | Risk Phrase | CAS No. | EINECS No. | USA TSCA | Canada DSL | Australia AICS | Korea ECL/MOE | Japan MITI/ENCS |
|---|---|---|---|---|---|---|---|---|---|
| 1 | Darocur 1173 | R36/37/38 | 7473-98-5 | 231-272-0 | Yes | Yes | Yes | | Yes |
| 2 | Omnirad 102 | | 68400-54-4 | | | | | | |
| 3 | Irgacure 184 | | 947-19-3 | 213-426-9 | Yes | Yes | Yes | | Yes |
| 4 | Irgacure 2959 | | 106797-53-9 | | | | | | |
| 5 | Omnirad 669 | | | | Yes | | | | |
| 6 | Esacure KIP 150 | | | | | | | | |
| 7 | Irgacure 127 | | 474510-57-1 | 400-600-6 | | | | | Yes |
| 8 | Esacure ONE | | | | Yes | | | | Yes |
| 9 | Irgacure PICS | | | | | | | | |
| 10 | Irgacure 907 | | 71868-10-5 | | Yes | Yes | Yes | | Yes |
| 11 | MMMP-3 | | | | Yes | | | | |
| 12 | Irgacure 369 | | 119313-12-1 | | Yes | Yes | Yes | | |
| 13 | BDMD-3 | | | | Yes | | | | |
| 14 | Irgacure 379 | | 119344-86-4 | | | | | | |
| 15 | Omnipol 910 | | | | | | | | |
| 16 | Irgacure 651 | R20/21/22, R51/53 | 24650-42-8 | 246-386-6 | Yes | Yes | Yes | Yes | Yes |
| 17 | Uvatone 8302 | | | | | | | | |
| 18 | DEAP | R38, R45, R20/21/22 | 6175-45-7 | | | | | | |
| 19 | i-propyl BE | | | | | | | | |

(continued on next page)

*Table A.11 (continued)* Registrations

| No. | Trade name | Risk Phrase | CAS No. | EINECS No. | USA TSCA | Canada DSL | Australia AICS | Korea ECL/MOE | Japan MITI/ENCS |
|---|---|---|---|---|---|---|---|---|---|
| 20 | n-butyl BE | R22 | 22499-12-3 | 245-039-6 | | | | | |
| 21 | i-butyl BE | R62, R51/53 | 75980-60-8 | 278-355-8 | | | | | |
| 22 | Lucirin TPO | R51/53 | 84434-11-7 | 282-810-6 | Yes | Yes | Yes | Yes | Yes |
| 23 | Lucirin TPO-L | | 162881-26-7 | | | | | | |
| 24 | Irgacure 819 | | 3584-23-4 | | | | | | |
| 27 | TAZ 104 | | | | | | | | |
| 28 | TAZ 110 | | | | | | | | |
| 31 | BMPS | R36/37/38 | 17025-47-7 | | Yes | | Yes | | |
| 35 | PDO | | 65894-76-0 | | | | Yes | | |
| 38 | BCIM | | 7189-82-4 | 230-555-6 | | No | | | |
| 42 | Irgacure 784 | | | | | | | | |
| 50 | Benzophenone | R52, R53, R36/37/38 | 119-61-9 | 204-337-6 | Yes | Yes | Yes | Yes | Yes |
| 51 | 4-methyl BP | R52, R53, R36/37/38 | 134-84-9 | 205-159-1 | Yes | Yes | Yes | | Yes |
| 52 | Kayacure MBP | | | | | | | | |
| 53 | Uvecryl P36 | | | | | | | | |
| 54 | Omnipol BP | | 515136-48-8 | | | | | | |
| 55 | Genopol BP-1 | | | | | | | | |
| 56 | Speedcure 7005 | tetra-<br>di- | 1003567-82-5<br>100357-16-1 | | | | | | |
| 58 | Speedcure PBZ | | 2128-93-0 | 218-345-2 | Yes | NDSL | No | | No |

| No. | Name | | CAS | | PMN | NDSL | | | |
|---|---|---|---|---|---|---|---|---|---|
| 60 | Speedcure BMS | | 83846-85-9 | 281-064-9 | PMN | NDSL | | Yes | Yes |
| 61 | Genocure MBB | | 606-28-0 | 210-112-3 | Yes | NDSL | Yes | | Yes |
| 62 | Esacure 1001M | | | | Yes | | | | |
| 63 | 2/4-ITX | | 75081-21-9 | 226-827-9 | Yes | Yes | Yes | Yes | No |
| 64 | 2-ITX | | 5495-84-1 | 281-065-4 | Yes | No | Yes | | No |
| | 4-ITX | | 83846-86-0 | 280-041-0 | Yes | Yes | | | |
| 65 | Kayacure DETX | | 82799-44-8 | 201-667-2 | Yes | No | Yes | Yes | Yes |
| 66 | Speedcure CTX | | 86-39-5 | ELINCS 7B | Yes | | | | Yes |
| 69 | Speedcure CPTX | | 142770-42-1 | | | | | | |
| 70 | Speedcure 7010 | | 1003567-83-6 | | | | | | |
| 71 | Omnipol TX | | 813452-37-8 | | | | | | |
| 72 | Genopol TX-1 | | | | | | | | |
| 73 | MBF | | 15206-55-0 | 239-263-3 | Yes | NDSL | Yes | | Yes |
| 74 | Irgacure 754 | | | | | | | | |
| 75 | CQ | | 10373-78-1 | | | | Yes | | |
| 76 | Benzil | R36/37/38 | 134-81-6 | | | | | | |
| 77 | EAQ | | 84-51-5 | 201-535-4 | Yes | Yes | Yes | | ENCS |
| 78 | Omnipol SZ | | | | | | | | |
| 114 | Speedcure DMB | R36/37/38 | 2208-05-1 | 218-630-1 | Yes | NDSL | | | |
| 115 | Speedcure EDB | | 10287-53-3 | 233-634-3 | Yes | Yes | Yes | | |
| 116 | Genocure EHA | R36/37/38 | 21245-02-3 | 244-289-3 | Yes | Yes | Yes | Yes | Yes |
| 117 | Kayacure DMBI | R36/37/38 | | | | | | Yes | Yes |
| 118 | Speedcure BEDB | | 67362-76-9 | 266-668-2 | Yes | NDSL | | | |

*(continued on next page)*

*Table A.11 (continued)* Registrations

| No. | Trade name | Risk Phrase | CAS No. | EINECS No. | USA TSCA | Canada DSL | Australia AICS | Korea ECL/MOE | Japan MITI/ENCS |
|-----|-----------|-------------|---------|------------|----------|------------|----------------|---------------|-----------------|
| 119 | Speedcure 7040 | tetra-<br>di- | 1003567-84-7<br>1003557-17-2 | | | | | | |
| 120 | Omnipol ASA | | | | | | | | |
| 121 | Omnipol ASE | | | | | | | | |
| 122 | Esacure A198 | | | | | | | | |
| 123 | Genopol AB-1 | | | | | | | | |
| 124 | Genopol RCX02 | | | | | | | | |
| 126 | EMK | R36/37/38 | 90-93-7 | 202-025-4 | Yes | | | | |
| 128 | NPG | | 103-01-5 | | | Yes | | | |
| 129 | LCV | R36/37/38 | 603-48-5 | 210-043-9 | | | | | |

# UV absorption curves

The absorption curves are set against a background of the output of a medium-pressure mercury lamp with the main lines at 250–260 nm, 313 nm, 365 nm, 404 nm and 436 nm showing clearly as sharp peaks. When other types of lamp are used, it is important to choose a photoinitiator that will absorb at the wavelength of the emitted light.

A photoinitiator will absorb UV energy at any point in its absorption pattern. It is not essential to match the peak absorption with a lamp output line; energy can be absorbed by the "tail" in absorption although this will lead to different cure patterns. Strong irradiation at the peak of absorbance will tend to give good surface cure. Irradiation at the "tail" in absorbance will tend to give good depth cure.

The absorption curves are listed numerically from the numbers in column 1 in the tables in Appendix A. The concentration of the photoinitiator is 0.02 g/dm$^3$ unless otherwise stated.

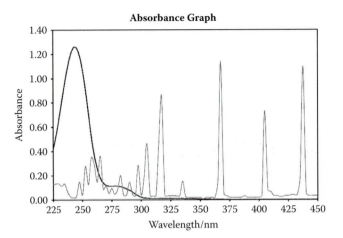

*Figure A.1* Darocur 1173 (1), 2,2-dimethyl-2-hydroxyacetophenone.

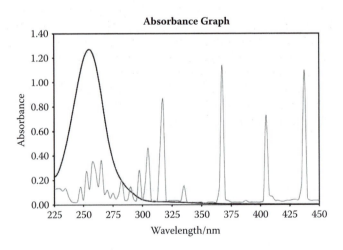

*Figure A.2* Omnirad 102 (2), 2,2-dimethyl-2-hydroxy-4'-(tert-butyl)acetophenone.

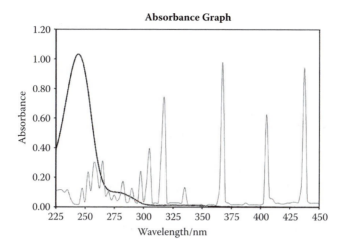

**Figure A.3** Irgacure 184 (3), 1-hydroxycyclohexyl phenyl ketone.

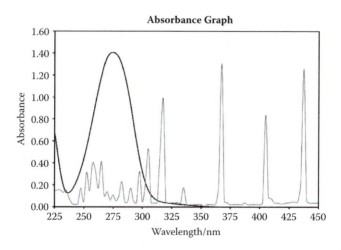

**Figure A.4** Irgacure 2959 (4), 2-hydroxy-4′-(2-hydroxyethoxy)-2-methylpropiophenone.

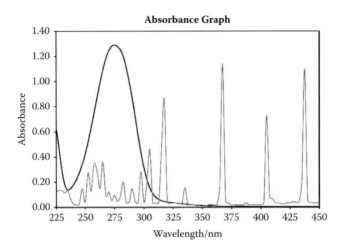

*Figure A.5* Omnirad 669 (5), 2-hydroxy-4'-(2-hydroxypropoxy)-2-methylpropiophenone.

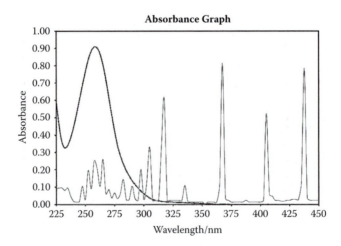

*Figure A.6* KIP 150 (6), polymeric.

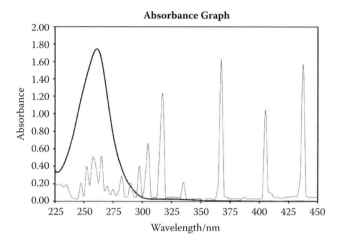

*Figure A.7* Irgacure 127 (7).

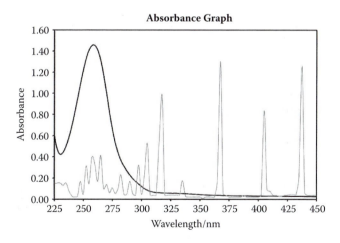

*Figure A.8* Esacure ONE (8).

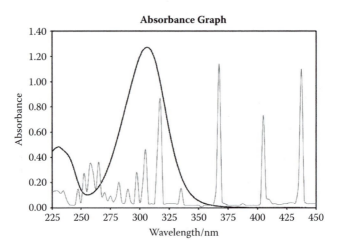

**Figure A.9** Irgacure 907 (10), 2-methyl-4′-(methylthio)-2-morpholinopropiophenone.

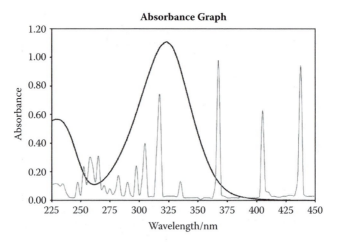

**Figure A.10** Irgacure 369 (12), 2-benzyl-2-(dimethylamino)-4′-(morpholino) butyrophenone.

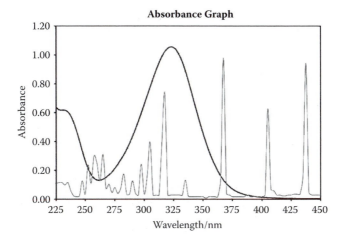

**Figure A.11** Irgacure 379 (14), 2-(4-methylbenzyl)-2-(dimethylamino)-4-morpholinobutyrophenone.

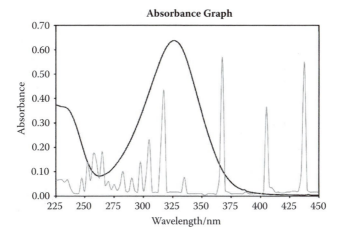

**Figure A.12** Omnipol 910 (15), polymeric.

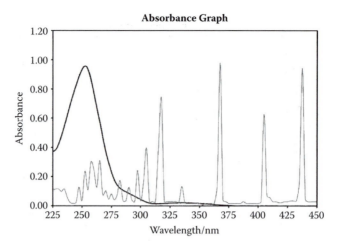

***Figure A.13*** Irgacure 651 (16), 2,2-dimethoxy-2-phenylacetophenone.

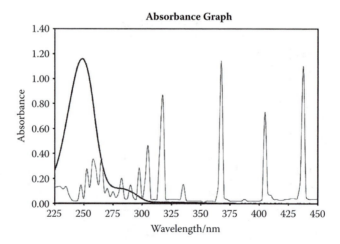

***Figure A.14*** DEAP, diethoxyacetophenone (18).

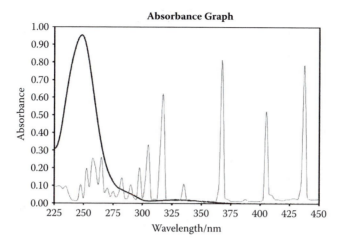

*Figure A.15* Benzoin isopropyl ether (19).

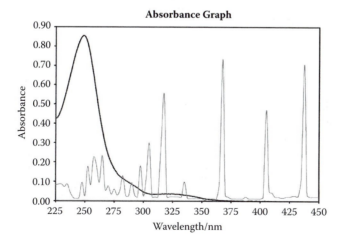

*Figure A.16* Esacure EB 3 (21), benzoin isobutyl ether.

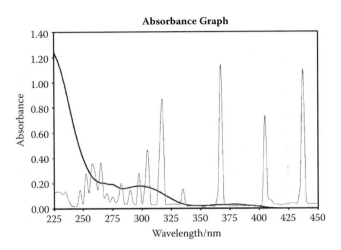

*Figure A.17* Lucirin TPO (22), (2,4,6-trimethylbenzoyl)-dihenylphosphine oxide.

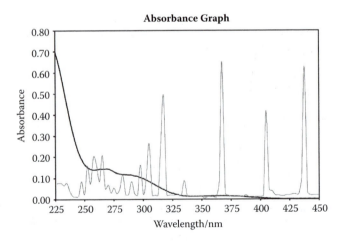

*Figure A.18* Lucirin TPO-L (23), Ethyl (2,4,6-trimethylbenzoyl)-phenylphosphine oxide.

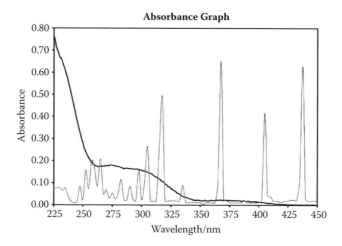

*Figure A.19* Irgacure 819 (24), Bis-(2,4,6-trimethylbenzoyl)-phenylphosphine oxide.

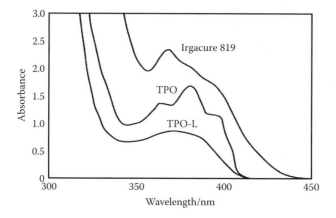

*Figure A.20* Phosphine oxides long wave absorption.

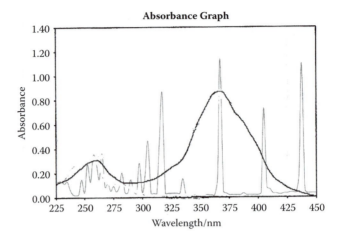

***Figure A.21*** QL Cure TAZ 110 (28), 2,4-bis(trichloromethyl)-6-p-methoxystyryl-
S-triazine.

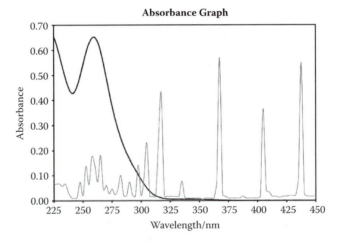

***Figure A.22*** Speedcure PDO (35), 1-phenyl-1,2-propanedione-2-(O-ethoxycarbonyl) oxime.

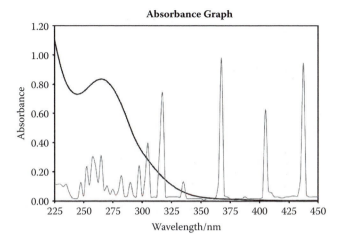

*Figure A.23* BCIM (38), 2,2′-bis(o-chlorophenyl)-4,4′,5,5′-tetraphenyl-1,2-biimidazole.

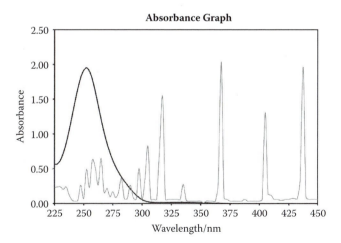

*Figure A.24* Benzophenone (50).

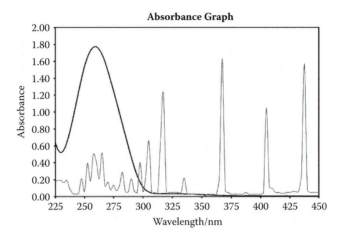

***Figure A.25*** Speedcure MBP (51), 4-methylbenzophenone.

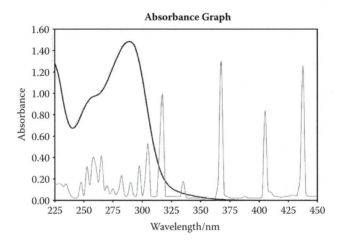

***Figure A.26*** Kayacure MBP (52), 3,3'-dimethyl-4-methoxybenzophenone.

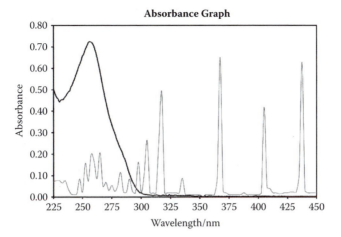

*Figure A.27* Uvecryl P36 (53).

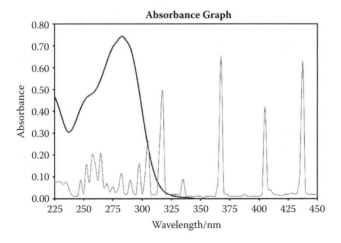

*Figure A.28* Omnipol BP (54), polymeric benzophenone.

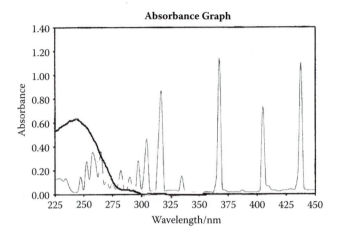

*Figure A.29*  Speedcure 7005 (56), polymeric benzophenone.

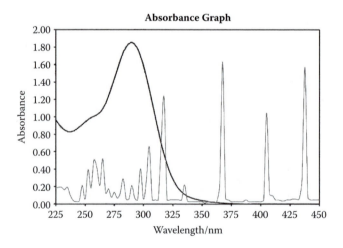

*Figure A.30*  Genocure PBZ (58), 4-phenylbenzophenone.

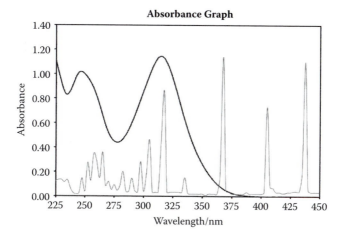

*Figure A.31* Speedcure BMS (60), 4-(4′-methylphenylthio)benzophenone.

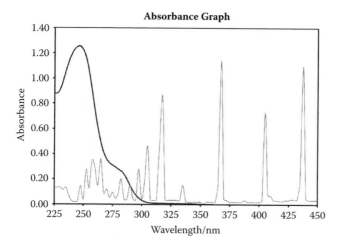

*Figure A.32* Genocure MBB (61), methyl 2-benzoylbenzoate.

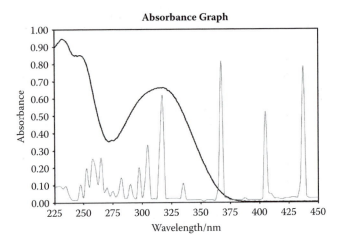

*Figure A.33*  Esacure 1001 M (62).

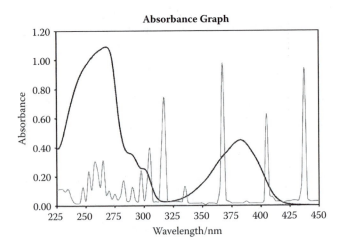

*Figure A.34*  Speedcure ITX (63), 2-, 4-isopropylthioxanthone, mixed isomers.

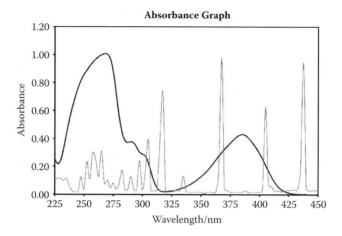

*Figure A.35* Kayacure DETX (65), 2,4-diethylthioxanthone.

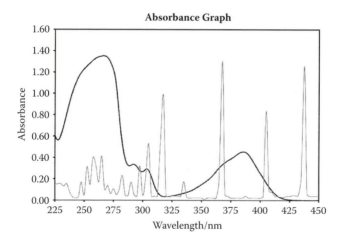

*Figure A.36* Speedcure CTX (66), 2-chlorothioxanthone.

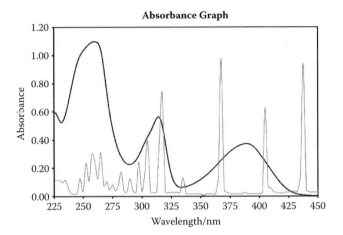

*Figure A.37* Speedcure CPTX (69), 1-chloro-4-propoxythioxanthone.

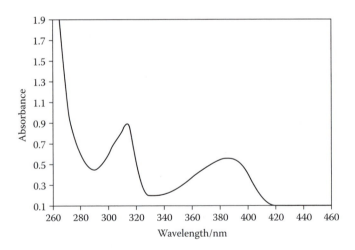

*Figure A.38* Speedcure 7010 (70), polymeric thioxanthone.

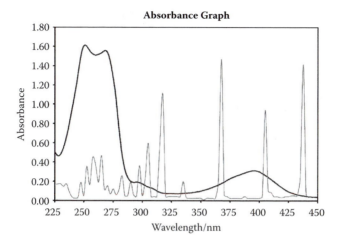

***Figure A.39*** Omnipol TX (71), polymeric thioxanthone.

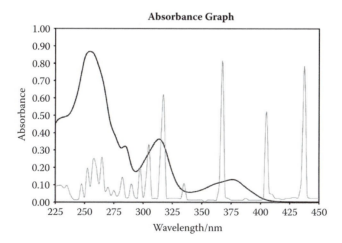

***Figure A.40*** Genopol TX-1 (72), polymeric thioxanthone.

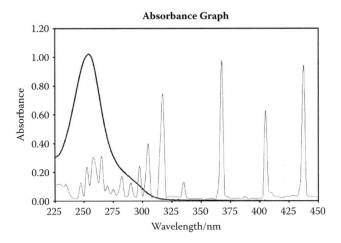

*Figure A.41*  Genocure MBF (73), methyl benzoylformate.

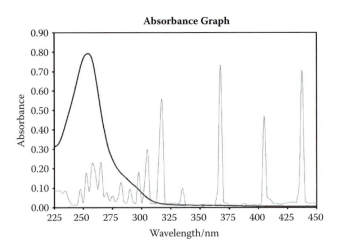

*Figure A.42*  Irgacure 754 (74), bis(benzoylformate) ester.

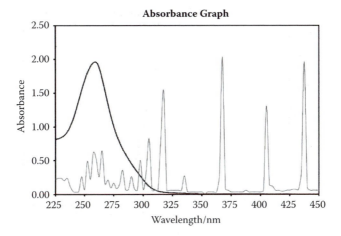

*Figure A.43* Benzil (76).

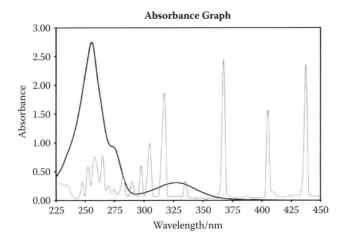

*Figure A.44* EAQ (77), 2-ethylanthraquinone.

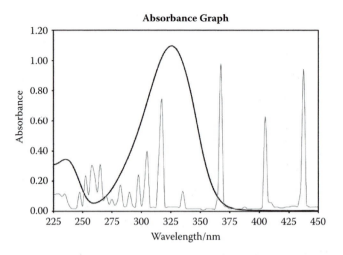

*Figure A.45* Omnipol SZ (78).

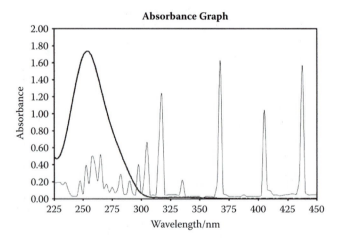

*Figure A.46* Omnirad BEM, a mix of BP and MBP (80).

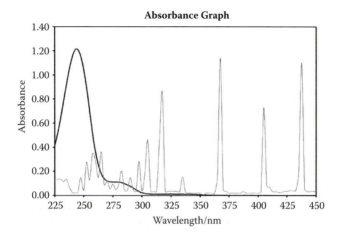

**Figure A.47** Irgacure 1000 (83), a mixture of 1173 and 184 (80/20).

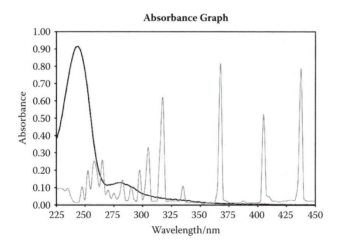

**Figure A.48** Irgacure 1700 (86), a mixture of 1173 and BAPO-1 (75/25).

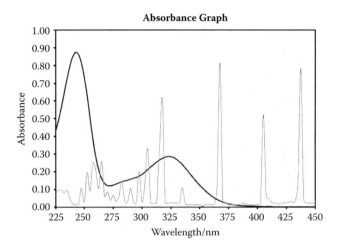

*Figure A.49* Irgacure 1800 (93), a mixture of 184 and BAPO-1 (75/25).

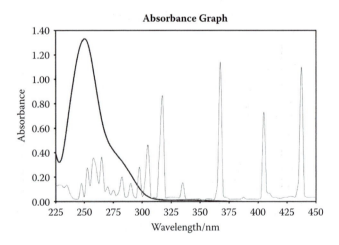

*Figure A.50* Esacure TZT (99), a mix of MBP and trimethyl BP.

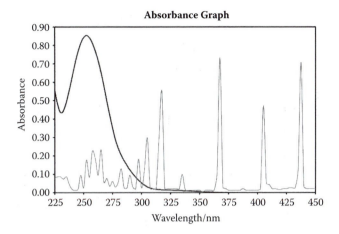

*Figure A.51* Esacure KIP 100F (101), a blend of KIP 150 and Darocur 1173.

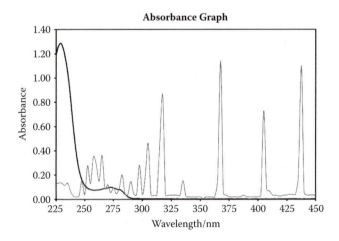

*Figure A.52* Speedcure DMB (114), 2-(dimethylamino)ethyl benzoate.

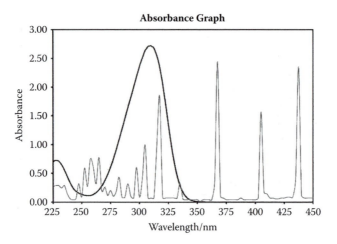

*Figure A.53* Speedcure EDB (115), ethyl 4-(dimethylamino)benzoate.

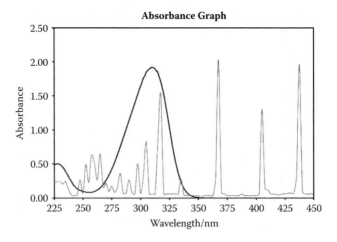

*Figure A.54* Speedcure EHA (116), 2-ethylhexyl 4-(dimethylamino)benzoate.

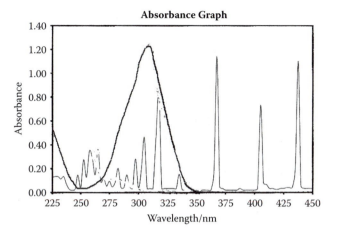

*Figure A.55* Speedcure 7040 (119), polymeric aminobenzoate.

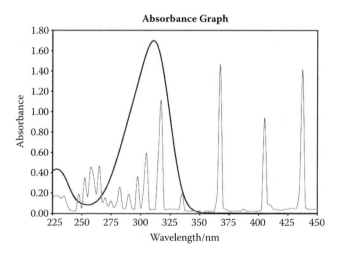

*Figure A.56* Omnipol ASA (120), polymeric aminobenzoate.

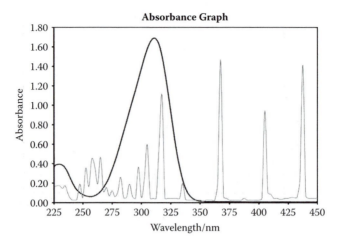

*Figure A.57*  Genopol AB-1 (123), polymeric aminobenzoate.

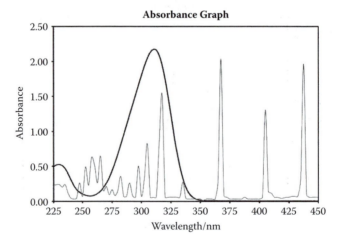

*Figure A.58*  Genopol RCX02 (124), polymeric aminobenzoate.

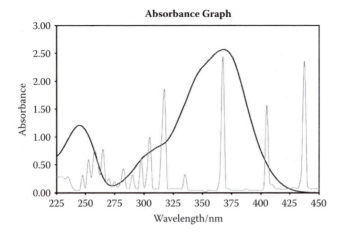

***Figure A.59*** Michler's ketone (125), MK, 4,4'-bis(dimethylamino)benzophenone.

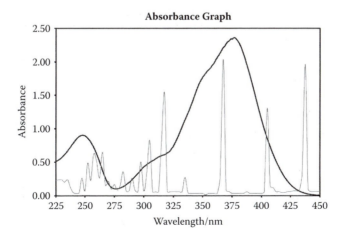

***Figure A.60*** Speedcure EMK (126), 4,4'-bis(diethylamino)benzophenone.

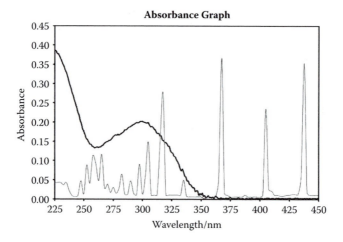

*Figure A.61* Cyracure UVI-6976 (140), sulphonium salt, $SbF_6^-$.

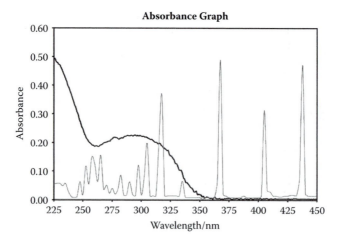

*Figure A.62* Cyracure UVI-6992 (141), sulphonium salt, $PF_6^-$.

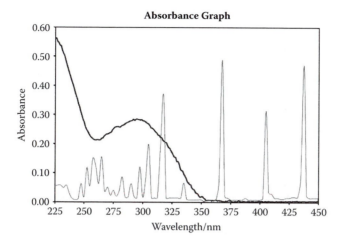

***Figure A.63*** Degacure K185 (142), sulphonium salt, $PF_6^-$.

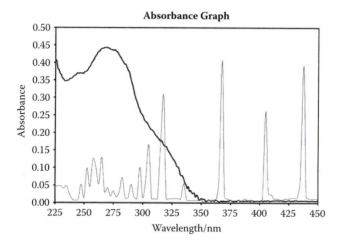

***Figure A.64*** SP-150 (145), sulphonium salt, $PF_6^-$.

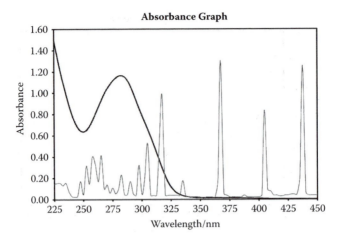

*Figure A.65* Omnicat 550 (147), sulphonium salt, $PF_6^-$.

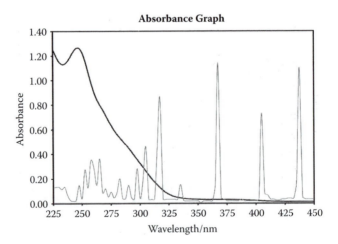

*Figure A.66* Omnicat 650 (149), polymeric sulphonium salt, $PF_6^-$.

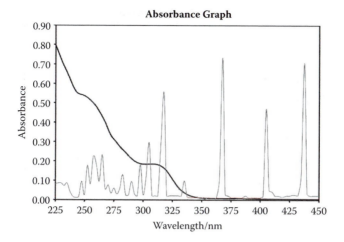

*Figure A.67* Esacure 1187 (150), sulphonium salt, $PF_6^-$.

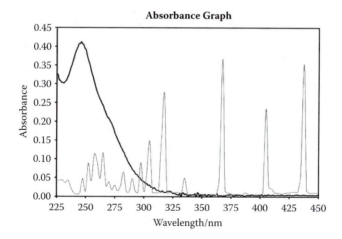

*Figure A.68* Sarcat CD-1012 (153), iodonium salt.

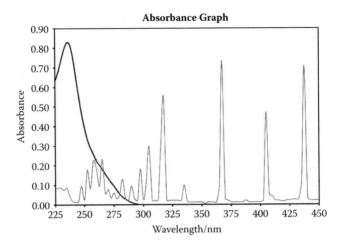

*Figure A.69*  Omnicat 440 (154), iodonium salt, $PF_6^-$.

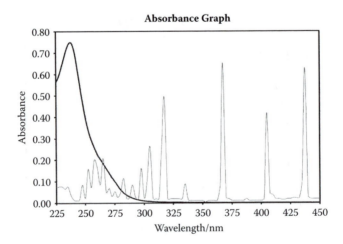

*Figure A.70*  Irgacure 250 (155), iodonium salt, $PF_6^-$.

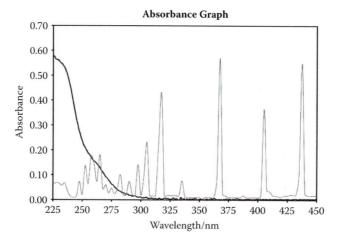

*Figure A.71* Rhodorsil 2074 (157), iodonium salt, borate anion.

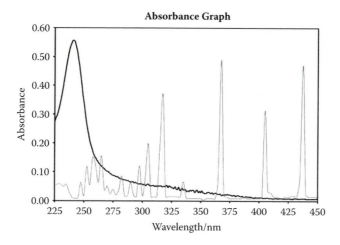

*Figure A.72* Irgacure 261 (158), ferrocenium salt, $PF_6^-$.

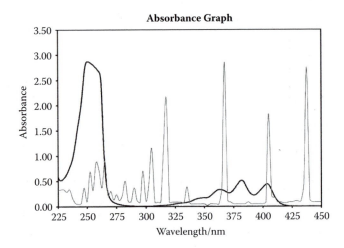

***Figure A.73*** Anthracure UVS-1331, DBA (160), anthracene sensitizer.

# appendix B

# Further information

Many textbooks about photoinitiators and UV curing are available that will provide much more information than this simple text, most including detailed academic studies of the photochemistry. Some of these are listed here.

1. *Photoinitiators for Free Radical, Cationic and Anionic Photopolymerisation*

   Wiley/SITA Series. Volume III. 2nd Edition
   J. V. Crivello and K. Dietliker. Edited by G. Bradley
   ISBN 0471 978922. 1998.

   A comprehensive and wide-ranging treatise on photoinitiators. This is one of several volumes in the Wiley/SITA series that covers all aspects of photopolymerization.

2. *Exploring the Science, Technology and Applications of UV and EB Curing*

   SITA Technology
   R. S. Davidson
   ISBN 0947 798412. 1999

   An excellent graduate introduction to many aspects of UV curing.

3. *Photogeneration of Reactive Species for UV Curing*

   Wiley
   C. Roffey
   ISBN 0-471-941778. 1998

   The mechanisms and photochemistry of initiators, monomers, oligomers, etc.

4. *Photoinitiation, Photopolymerisation and Photocuring: Fundamentals and Applications*

Hanser Publishers
J. P. Fouassier
ISBN 1 56990 1465. 1995

A comprehensive text of both academic studies and practical application.

5. *A Compilation of Photoinitiators Commercially Available for UV Today*

SITA Technology
K. Dietliker
ISBN 9477 98676. 2002

6. *Radiation Curing in Polymer Science and Technology. Volume II. Photo-initiating Systems*

Elsevier Applied Science
Edited by J. P. Fouassier and J. F. Rabek
ISBN 1-85166-933-7. 1993

Both academic and practical, if a little dated.

7. *UV and EB Curing Formulations for Printing Inks, Coatings and Paints. 2nd Impression*

SITA Technology
R. Holman and P. Oldring.
ISBN 0-947798-02-1. 1988

8. *Test Methods for UV and EB Curable Systems*

SITA Technology
C. Lowe and P. K. T. Oldring
ISBN 0-94779807-2. 1996

9. *UV Coatings. Basics, Recent Developments and New Applications*

Elsevier
R. Schwalm (BASF)
ISBN 978-0-444-52979-4. 2007

An excellent, up-to-date coverage of the wide variety of UV coatings, with practical advice on many aspects.

## Conference Proceedings

Conference proceedings, a good source of practical advice, generally follow the biennial conferences of the RadTech Organization, held separately on several continents.

RadTech Australia
6 Talbot Street
Peakhurst, NSW 2210
Australia
Tel. +612 9584 8044
Fax. +612 9584 8055
Email alphauve@sunink.com

RadTech China
PO Box 6
Beijing University of Chemical Technology
Beijing, 100029
P.R. China
Tel. +86 10 6441 5732
Fax. +86 10 6441 5715
Email org@radtechchina.com
www.radtechchina.com

RadTech Europe
PO Box 85612
NL-2508 CH
The Hague, the Netherlands
Tel. +31 (0)70 312 3916
Fax. +31 (0)70 363 6348
Email Mail@radtech-europe.com
www.radtech-europe.com

RadTech Japan
401 Soshu Building
4-40-13, Takadanobaba
Shinjyuku-ku
Tokyo 169-0075
Japan
Tel. +81 03 3360 0135
Fax. +81 03 3360 2270
Email kichimur@res.titech.ac.jp

RadTech Malaysia
Bangi
Malaysia Institute for Nuclear Technology Research
430000 Kajang
Malaysia
Tel. +603 828 2912
Fax. +603 820 2968
Email radtech@mint.gov.my

RadTech South America (Brazil)
Av. Prof. Lineu Prestes
2.242–Predio do CTR–sala
São Paulo–05508-000
Cidade Universitaria, Brazil
Tel/Fax. +55 11 3034 0078
Email atbcr@apin.brn
www.atbcr.com.br

The Paint Research Association is an excellent base for study, training, and advice about UV matters. It has also followed the biennial trend up to 2006 with its Radcure Conferences in the UK. Proceedings can be found at

Paint Research Association
14 Castle Mews, High Street
Hampton, Middlesex
TW12 2NP, UK
Tel. +44 (0)20 8487 0800.
Fax. +44 ()20 8487 0801
www.pra-world.com

# *Index*